Handbook of Railway Engineering and Security

Handbook of Railway Engineering and Security

Edited by **Marshall Roy**

LANRYE
INTERNATIONAL

New Jersey

Published by Clanrye International,
55 Van Reypen Street,
Jersey City, NJ 07306, USA
www.clanryeinternational.com

Handbook of Railway Engineering and Security
Edited by Marshall Roy

International Standard Book Number: 978-1-63240-287-5 (Hardback)

Printed in the United States of America.

Contents

Preface

Ever since the first ever train was used to carry coal from a mine in Shropshire (England, 1600), the technology of railway transportation has never looked back. It has only evolved and developed and remains one of the most important developments in the history of mankind even in today's age. The biggest invention in this field was the development of steam locomotive, but it took another two hundred years for commercial rail travel to practically begin. The railway systems of present day are much more complicated than they earlier used to be. This book is devoted to parameters monitoring in railway construction for safety and reliability purposes. This book provides a technical guide for those interested in learning about railway engineering and security.

This book has been the outcome of endless efforts put in by authors and researchers on various issues and topics within the field. The book is a comprehensive collection of significant researches that are addressed in a variety of chapters. It will surely enhance the knowledge of the field among readers across the globe.

It is indeed an immense pleasure to thank our researchers and authors for their efforts to submit their piece of writing before the deadlines. Finally in the end, I would like to thank my family and colleagues who have been a great source of inspiration and support.

Editor

Signalling, Security and Infrastructure Construction in Railway

Criteria for Improving the Embankment-Structure Transition Design in Railway Lines

Inmculada Gallego, Santos Sánchez-Cambronero and Ana Rivas
University of Castilla-La Mancha
Spain

1. Introduction

In the design of a railroad track there are some situations in which the introduction of a structure in the track is needed, for example a bridge, a viaduct or a pontoon. This circumstance is even more frequent in the High Speed lines, since the design criteria, fundamentally slopes and radius, are stricter than those for conventional lines. The introduction of a structure determines the appearance of a point with an abrupt change in the vertical stiffness from a track cross section to another.

The experience has shown that these transition zones between embankment and structure are the source of many problems (related to safety, passengers' comfort, maintenance expenses, etc.), causing differential settlements among adjacent track cross sections and originating which is known as "dip" (European Rail Research Institute, ERRI., 1999)

In order to diminish this unfavorable effect, the well known "technical blocks" are designed in a length determined between the structure and the embankment of access to this one. However, in spite of this structural disposition, it has not been found yet any design solution that notably reduces the track geometrical quality defects that have been observed in the mentioned zones. This is an important issue, because they produce a relevant increase in the maintenance expenses of the High Speed lines and they affect the availability of the track (Gallego, López, Ubalde, & Texeira, 2005).

2. Theoretical foundation of embankment-structure transition behavior

The wheel load transmitted by a train to the track does not correspond to the static load; instead, random dynamic overloads appear due to the sprung and un-sprung masses. Among the great amount of existing formulations relative to these overloads, there is an outstanding contribution made by Prud 'Homme (Prud`Homme, 1970) according to the expression (1)

$$\sigma\left(\Delta Q_{NS}\right) = 0.45 \frac{V}{100} b \sqrt{m K \varphi(\varepsilon)} \,, \tag{1}$$

where: $\sigma(\Delta Q_{NS})$ is the standard deviation of the dynamic overloads due to the un-sprung masses of the material; V is the running speed of the vehicle; b is a variable related to the track defects and to the vehicle defects; m is the un-sprung mass of the vehicle; k is the vertical track stiffness; $\varphi(\varepsilon)$ is Damping of the track.

Expression (1) introduces a new criterion to reduce the mutual aggressiveness between track and vehicle. From that it is deduced the importance of having a low value of the vertical stiffness of the track (K) and of the un-sprung mass of the vehicle (m) to avoid increasing the dynamic overloads due to the un-sprung masses. This influence is a more relevant fact in high speed trains.

In addition, it is not only the stiffness value which determines the dynamic overloads. They are also influenced by the variation of the stiffness value that might exist from one sleeper to another.

The first studies carried out to deal with this problem were carried out by Amielin, 1974, later on in the eighties they stand out those by Lopez, 1983, Hettle, 1986, Hunt, 1997, Esveld, 2001, López A. , 2001 or Teixeira, 2003 shown that as the difference between the stiffness values of two consecutive sleepers increases, the reaction on the sleepers increases, thereby increasing the load on the sleeper. On the other hand, next to these increments of stress, the experience has proved that some differential settlements are originated. As a result of these two factors hanging sleepers can be developed that in turn increase the stress on the ballast. In order to avoid this deterioration experimented in the transitions, these sections are built the well known "technical blocks", whose aim is achieving a gradual increase in the stiffness from one sleeper to the following one, as we reach the structure.

Now, it is interesting to know: How are these designs? What criteria are used to define them? To answer these questions a revision of the designs used by the different European Railway Administrations has been made. Five types of the most used measures have been identified. They are enumerated next, being the first one the most frequently employed:

- Backfill behind the abutment either with materials of a high compression level or granular material treated with cement.
- Use of a transition slab built with reinforced concrete or another material.
- Introduction of horizontal layers on a track formation of different materials.
- Use of geosynthetics to achieve an abutment reinforced backfill.
- Treatment of the track bed and sub-ballast with cement.

Along with these measures, they have been also identified a variety of track formation materials behind the abutment. There are three types that stand out, just as it is schematized in Fig 1. The first two types are the more frequently used, and with regard to that work they will be called slope type PA and slope type PB.

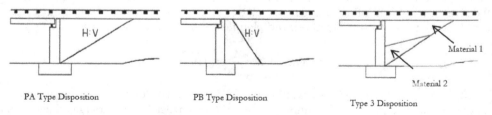

PA Type Disposition   PB Type Disposition   Type 3 Disposition

Fig. 1. Types of dispositions of the backfill behind the abutment in embankment-structure transitions.

This revision of the solutions employed, shows a lack of homogeneity of the design criteria: each Railway Administration uses different designs for the longitudinal sections of the embankment-structure transitions.

3. Motivation of the chapter

The previous analysis shows a lack of homogeneity in the design criteria. Besides, it must be added what the experience has pointed out, a remarkable deterioration of the quality in the transitions. These two facts, lead to think about two reflections: a) it does not exist a precise knowledge of the behavior of transitions, and b), the current designs have certain limitations since they are unable to reduce in a remarkable way the deterioration experimented in these areas. The first reflection invites us to deepen in the knowledge of the deterioration process and the second one induces us to consider two more aims: 1) introducing new design criteria and 2) giving an analytical basis to the already existent, to overcome some of those limitations. In order to achieve those aims a numerical modelization of the embankment-structure transition is shown in this chapter (Gallego & López, , 2009)

Therefore, within the scope of this work it is sought to carry out a model of finite elements which simulates the behavior of embankment-structures transitions. This model will enable to quantify the vertical stiffness of the track according to the type of disposition of the materials which the transition is carried out with (PA and PB, Fig 1) as well as the geotechnical characteristics of these materials (See Tab 1 and Tab 2).

Material	E (N/m^2)	υ	c (N/m^2)	ϕ (^{0})	ρ (N/m^3)
Rail steel	2.1×10^{11}	0.3	-	-	7.5 10^4
Elastic bearing	2.952×10^8	-	-	-	-
Sleeper element 1	8.01×10^{10}	0.25	-	-	-
Sleeper element 2	5.02×10^{10}	0.25	-	-	-
Sleeper element 3	3.68×10^{10}	0.25	-	-	-
Ballast	1.3×10^8	0.2	0	45	1.9×10^4
Sub-ballast	1.2×10^8	0.3	0	45	1.9×10^4
Track bed	8.10^7	0.3	0	35	2×10^4
Material QS1	1.25×10^7	0.4	15000	10	2×10^4
Material QS2	2.50×10^7	0.3	10000	20	2×10^4
Material QS3	8×10^7	0.3	0	35	2×10^4
Cement-treated granular material	1.6×10^8	0.25			2.3×10^4
Rock	3×10^9	0.2			2.7×10^4

Table 1. Values of geotechnical parameters considered in the models

GEOMETRIC PARAMETERS		GEOTECHNICAL PARAMETERS		Name of case studies H embankment (H7)-
Type of design (Fig 1)	Slope value (H:V)	Type material of transition (type1/type2)	Original ground	Disposition – value of the slope- material 1- material2-original ground
PA	1:1	QS2/QS3	QS1	H7PA11QS2QS3QS1
PB	3:1	QS2/MGT* QS3/MGT*	QS2 QS3 ROCA	

*MGT = Cement-treated granular material

Table 2. Geometric and geotechnical parameter considered in the model

4. Description of the numerical model and the work assumptions

In order to quantify the vertical track stiffness value and the incidence that the disposition and type of material of the track formation have on it, it is considered more accurate to analyze the track as a whole system. So as to carry out this analysis, the most appropriate approach is to apply the finite elements method. This method enables the numerical simulation of different materials and diverse boundary conditions, facilitating the study of the interaction among the different elements that compose the railway superstructure and infrastructure.

The employment of the finite elements method is accurate to evaluate the global behavior of the track structure, but it is very limited to quantify the efforts in the ballast: the inter-granular stress is very different from stress and strain assumed in a continuous medium. However, the use of a discrete elements model, or a mixed one, finite and discrete elements, implies a tremendous computational cost and an enormous complexity.

In the Railway field, the finite elements method has been used by authors as López A. , 1977, Sauvage & Larible, 1982, Profillidis, 1983, Sahu, Rao, & Yudhbir., 1999 Mira, Férnández, Pastor, Nasarre, & Carrillo, 2000, among others. Some studies developed in the eighties stand out, such as Profillidis, 1983, since the results of those works were integrated by the Committee D-117 of the ORE in the Record (Comité D-117 (ORE), 1983). The sizing graphics of the track bearing structure are collected in this work. Together with the previous works, it is also worth mentioning those carried out by the "Ministerio de Fomento Español", which have been the basis for making some recommendations for the Railways track construction.

In order to generate the model proposed in this chapter, the contributions collected in the different models of Railways track formations carried out so far and enunciated previously have been taken into consideration.

4.1 Description of the analyzed domain

The length of an embankment-structure transition depends on the type of structure and the height of the access embankment. In the case of embankment heights around 15 meters, the technical block can reach lengths of up to 85 meters (Fig 2). A 3D Model for such that length require an extremely powerful software and hardware, implying a huge computational cost due to the long time calculation that would be needed.

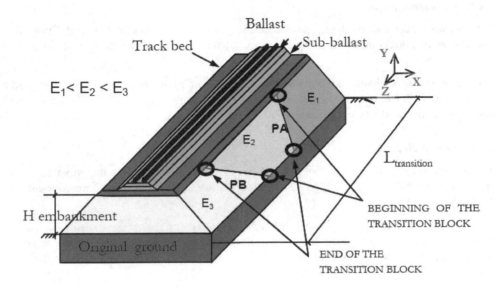

Fig. 2. General schematic of an embankment–structure. The transition is the zone adjacent to the structure (see Gallego I.,2006)

Suitable accuracy can be achieved by modeling only a transition section. However, the track section chosen to be modeled should be such that possesses the fundamental characteristics of a transition. However, the track section chosen for modeling should have the fundamental characteristics of a transition. For both slope types considered (PA and PB), these characteristics occur when the material changes from one material to another, either at the beginning or end of the slope (see Fig 2). Thus, the sections to be modeled are those shown within the rectangles in Fig 3.

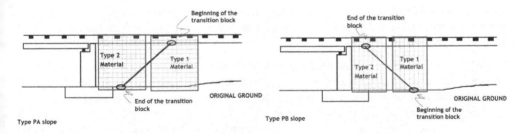

Fig. 3. Schematic of the beginning and end of the technical block in the slope of type PB and PA

4.2 Geometry of the analyzed domain

The directions considered for the model are the following ones: the axis x indicates the sleeper direction, the axis y the vertical and the axis z the rail direction (see Fig 2).

4.2.1 Transverse section

The measures adopted for the transversal section are the ones used in the High Speed line Madrid-Sevilla, but considering the track as simple and applying symmetry with respect to the plane $x = 0$.

The slope of the embankment varies with the type of material from which it is made; between H/V=3/2 and 1/2. The value 3/2 has been adopted in the model, because it was employed in the Madrid–Seville line.

4.2.2 Vertical direction

In the vertical direction (axis y) not only all the elements that compose the super structure are considered but also the, sub-ballast , the formation layer, 7m for the embankment and 3m for the original ground.

Values of 30, 25, and 60 cm, respectively were adopted for the thicknesses of the ballast under a sleeper, the sub-ballast the formation layer. These values are the same values as those applied in the transversal section of the High Speed line between Madrid and Seville.

4.2.3 Longitudinal direction

For domain analysis, the load applied on a sleeper is transmitted to the adjacent sleepers through some transmission coefficients. These coefficients clearly decrease as the distance from the point of load application increases; the coefficient is only 7 per cent in the third sleeper, when the first sleeper is defined as the one on which the load is applied (Comité D-117 (ORE), 1983)

To determine the real value of the settlement of the head of the rail when the load acts on it, one must not only solve for that load but also consider the history of the previous loads that have affected that sleeper. The load applied on a sleeper transmits it to the two adjacent sleepers. In order to observe the behavior of two consecutives sleepers with different stiffness under a load, while taking into account their load history also, four successively loaded sleepers (T5, T6, T7, and T8) were considered in order to analyze the behavior of sleepers T7 and T8.

In 1983, a test in Derby showed that important phenomena are apparent up to the fourth sleeper from the one loaded (Comité D-117 (ORE), 1983). Therefore, four unloaded sleepers were introduced at both ends of the model. This set-up avoided artefacts and yet included a substantial number of sleepers onto which the load could be applied. This approach required consideration of a transition sector comprising 12 sleepers, leading to a model system with a total length of 7.20 m.

4.3 Modelling rail track, elastic bearing, and sleeper section

In order to model the rail track, its resistance to bending was simulated in the most accurate way possible (Fig 4), which is why the inertia of the modelled rail must be equal to that of the real rail.

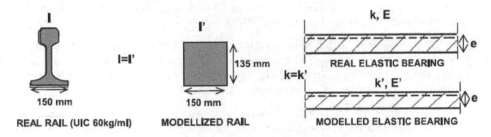

Fig. 4. Model of the rail and elastic bearings carried out in this study

The model also sought to make the vertical stiffness equal for all of the elastic bearings (see Fig 4). The vertical dimension and the modulus of elasticity were fixed so that the vertical stiffness of the element coincided with the stiffness of the elastic bearing provided by the manufacturer. For the high-speed Madrid–Seville line, the elastic bearings have a stiffness of nearly 500kN/mm (López A. , 2001)

Because the sleeper section is not constant along its entire length, the dimensions of its most representative section were used for the sleeper model elements. For each element (See Fig 5), the modeled flexural stiffness must be equal to the real flexural stiffness, as follows

$$E_{model}I_{model} = E_{real}I_{real} \qquad (2)$$

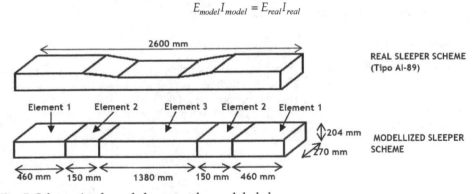

Fig. 5. Schematic of a real sleeper and a modeled sleeper

To obtain homogeneity in the calculations, the elements of the model had to have a constant width, which is not the case in the real sleeper. Thus, the model width must be considered to be an average width. This average width must be such that the load bearing surface in the model is equal to that in reality.

4.4 Sleeper–ballast contact

The sleeper–ballast contact zones contain a high concentration of strains. This local phenomenon requires refining of the mesh used to model these zones. However, applying this procedure is sometimes impossible because of the computational resources and model complexity needed. The most common alternative to modeling the contact zones is to use bounded degrees of freedom. In fact, this solution was adopted by ORE Committee D-117 and was used by the Railway Track Formations Project in its recommendations on railway track construction (Ministerio de fomento, 1999).

The use of bounded degrees of freedom requires the introduction of different nodes for each material at the contact surface. These nodes must move equivalently in the direction perpendicular to the contact plane (see Fig. 6). However, these nodes can move at different values in the directions parallel to the contact plane.

This solution is effective because it solves the tensional discontinuities that appear at the interface between two materials that differ significantly in their stiffness. In this model, bounded degrees of freedom were used at the sleeper–ballast contacts.

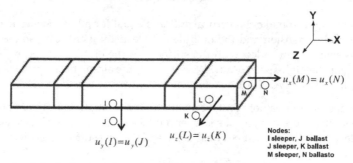

Fig. 6. Schematic of the ballast-sleeper contact

4.5 Boundary conditions

The model in this study differs from many existing models of railway track construction (Gallego, López, Ubalde, & Texeira, 2005), in which all vertical planes are constrained in all directions. In the Supertrack project (European comunity, 2005) and in this work, the planes that shape the slopes of embankments are left completely free, with no restrictions. In particular, the boundary conditions used here are as follows (see Fig. 7):

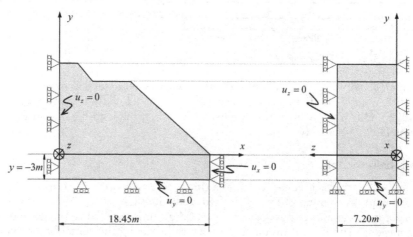

Fig. 7. Boundary conditions

- In the vertical plans limits of the model, z=0 and z=7.20m, the boundary condition adopted is to impose the nullity of movement in the perpendicular direction to these plans (u_z=0).
- In the vertical plans limits of the model, x=0 and x=18.45m, the boundary condition is, like the previous one, to impose the nullity of movement in the perpendicular direction to these plans (u_x=0).
- In the horizontal inferior plan of the model y=-3m, the condition to be imposed is the null vertical displacement (u_y=0).

4.6 Material constitutive model

An elastic, isotropic, and linear model was used to develop a mechanic model of the rail tracks, elastic bearings, sleepers, and the granular material processed with cement. In contrast, a perfect plastic model, i.e. the Drucker–Prager model (Oliver & Arlet, 2000), was used to model the rest of the materials, including the ballast, sub-ballast, track bed, embankment fill, and original ground. Finite strain was used to simulate the kinematics of the continuous medium.

Granular material treated with cement is considered to behave elastically, at least until it reaches a substantial percentage of its stress limit; one can assume its modulus of elasticity to remain essentially constant under normal stress.

To model the embankment material on which the track is laid, a perfect elasto-plastic behavior was assumed. This assumption implies that reloading occurs in the same way as downloading, and that the material experiences no hardening (hardening parameter $H=0$).

For modeling of the yield surfaces, the most accurate approach is to use a model dependent on hydrostatic pressure. These models are the Drucker–Prager and Mohr–Coulomb models, which limit the material behavior for states of hydrostatic stress (in traction). For the present study, the Drucker–Prager model was selected because it has been used in several elasto-plastic models used to design railway projects, and it has been validated by ORE Committee D-171.

The principle formulated by Drucker and Prager in 1952 includes the influence of pressure through the first invariant of the stress tensor I_1 and the internal friction angle φ. It also depends on the second invariant J_2 of the deviatoric stress tensor, as well as on two parameters: the friction angle among particles φ and the cohesion c. This criterion is expressed by means of the principal stress invariant and J_2 (the second invariant of deviatoric stress), as follows:

$$F\left(I_1, J_2, c; \phi\right) = \bar{\alpha}(\phi) I_1 + \sqrt{J_2} - \bar{K}(\kappa, \phi) = 0 , \tag{3}$$

where

$$\bar{K}(\kappa, \phi) = 6\, c(\kappa) \cos\phi \,/\, (3\sqrt{3} + \sqrt{3}\,\mathrm{sen}\,\phi) \tag{4}$$

and

$$\bar{\alpha}(\phi) = 2\,\mathrm{sen}\,\phi \,/\, (3\sqrt{3} + \sqrt{3}\,\mathrm{sen}\,\phi) \tag{5}$$

When the function that defines the yield surface (3) is represented in the main stress space, a cone is obtained, the axis of which is the hydrostatic axis (see Fig 8).

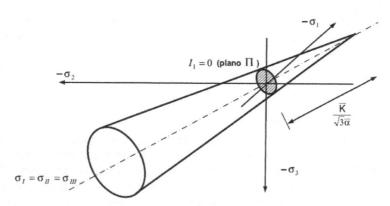

Fig. 8. Schematic of yield surfaces Drucker–Prager

Of the elements used, a quadratic 'brick' element of 20 nodes was selected; this is the most common type of element in three-dimensional models used to design railway tracks. The final meshes are shown in Fig 9.

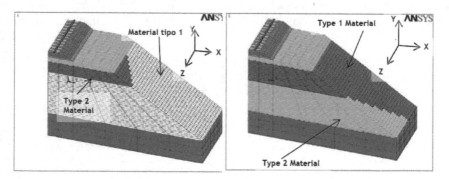

Fig. 9. Finite-element model for transitions with slopes of type PA31 and PB11

4.7 Hypothesis adopted

The load application must be carried out in several stages. In the first stage, only the material's own weight is considered until reaching the stress balance, while at later stages the loads due to the train are also taken into account. The stresses and displacements of interest are the ones that correspond to the application of the train loads; therefore, they can be calculated from the difference between the totals obtained after applying the train loads to the first stage.

Here, it was convenient to apply four load states due to the train passage, matching each state to the application of the static load per wheel in the four central sleepers of the model: T5, T6, T7, and T8 (See Fig 10).

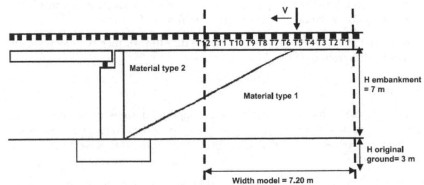

Load state 1 (LS1): only own weight

Load state 2 (LS2): static load on sleeper T5

Load state 3 (LS3): T5 sleeper download and T6 load

Load state 4 (LS4): T6 sleeper download T6 and T7 load

Load state 5 (LS5): T7 sleeper download and T8 load

Fig. 10. Schematic of the different load states

To simulate the constructive process of an embankment and to ensure convergence of the solution, the first load state (only the material weight) was divided into 250 substeps of the gravitational load application and 50 balance iterations for each one. For railway loads, 15 sub-steps proved to be sufficient to achieve convergence. The program used is ANSYS Structural, which enables nonlinear analysis.

To carry out the calculations, the load of only one axle was considered while assuming the effects of the remaining axles to be negligible.

The increased value of the dynamic overloads was calculated using Prud'Homme's formulation, with 1mm used as the value for b and the speed set to be 300 km/h. With this approach, the value of a dynamic overload was obtained according to the dynamic stiffness, which is unknown because its calculation requires knowing the value of the point dynamic load on the track. It is customary to carry out railway calculations assuming that the dynamic stiffness has a value similar to that of the static stiffness; this assumption has been validated by experience and confirmed by calculations (Ministerio de fomento, 1999). Thus, this assumption was also made in the present study, and numerical tests were used to justify this choice.

5. Results from the model: Presentation and critical analysis

This modeling study was carried out in two phases with different aims. In the first phase, different case studies were analyzed numerically, and the results were studied. These results then determined whether to carry out more case studies or proceed with the second phase. The main objective of the first phase was to observe how changing from a less rigid material to a more rigid one would affect the vertical stiffness of the track.

In the second phase, variations caused by the different stiffness values were analyzed for all the case studies from the first phase.

5.1 First phase

In the first phase, the main differences among the designs of the embankment–structure transitions are the types of materials used and the construction design. This justifies defining two types of case studies based on different geotechnical and geometrical parameters.

The geometrical parameters include the type and grade of slope; the latter is defined, e.g., as 1 : 1 and 3 : 2 ($H : V$, horizontal and vertical). The geotechnical parameters are the modulus of elasticity of the materials that compose the embankment fill and the original ground. For the modulus of elasticity, a range of values was used, matching those adopted in the numerical model presented by ORE Committee D-117. The values coincide with the lower limit values corresponding to the material types QS1, QS2, QS3, and rock (see Tab 1).

To fill the embankment, granular material processed with cement (MGT) was added to the model, since this material is so frequently used. Combining all these values with the technical block and the four load states already described, yielded a total of 48 case studies (see Tab 2). The modeling results for the case studies are shown in Tab 3.

To make additional comparisons, the stiffness was calculated for cases in which the fill corresponded to conventional embankments made with the same type of material (see Tab 4).

1. It is useful to apply stiffness values not only at the beginning but also at the end of the technical block; therefore, the cases corresponding to the ends of the technical blocks were calculated. Since the calculated stiffness values in the first 48 cases were similar for original ground QS1 and QS2, and for QS3 and rock, it was sufficient to calculate the cases corresponding to QS2 and QS3, thereby reducing the number of cases from 48 to 24 (see Tab 5).

5.2 Second phase

The second phase involved analyzing the results of the first phase. The criteria were to limit the following:

- The upper value of the vertical stiffness
- The lower value of the vertical stiffness
- The value of the longitudinal variation

TRANSITION TYPE	ORIGINAL GROUND QS1	ORIGINAL GROUND QS2	ORIGINAL GROUND QS3	ORIGINAL GROUND ROCK
H7PA11QS2QS3	11.242	17.227	37.206	42.752
H7PB11QS2QS3	18.847	19.957	36.054	39.325
H7PA31QS2QS3	10.800	16.380	34.139	39.744
H7PB31QS2QS3	14.015	19.296	34.519	38.512
H7PA11QS2MGT	15.380	25.499	43.859	47.795
H7PB11QS2MGT	28.721	32.644	37.657	40.000
H7PA31QS2MGT	13.182	18.275	36.549	42.077
H7PB31QS2MGT	21.524	25.961	35.777	38.592
H7PA11QS3MGT	32.193	53.563	71.692	75.374
H7PB11QS3MGT	54.186	59.936	69.294	70.473
H7PA31QS3MGT	29.354	51.068	70.875	74.650
H7PB31QS3MGT	48.668	58.069	70.340	73.444

Table 3. Values of vertical static track stiffness at the beginning of transition K_I (kN/mm) for all cases studied

EMBANKMENT MATERIAL	ORIGINAL GROUND QS1	ORIGINAL GROUND QS2	ORIGINAL GROUND QS3	ORIGINAL GROUND ROCK
FULL QS2 ($K(QS2)$)	10.186	15.717	33.226	38.433
FULL QS3 ($K(QS3)$)	28.899	50.929	70.606	74.027
FULL MGT ($K(MGT)$)	67.782	73.676	83.214	87.800

Table 4. Values of vertical static track stiffness K for conventional embankments (kN/mm)

TRANSITION TYPE	ORIGINAL GROUND QS2		ORIGINAL GROUND QS3	
	BEGINNING (K_I)	END (K_F)	BEGINNING (K_I)	END (K_F)
H7PA11QS2QS3	17.227	37.581	37.206	61.927
H7PB11QS2QS3	19.957	47.071	36.054	64.276
H7PA31QS2QS3	16.380	41.982	34.139	65.175
H7PB31QS2QS3	19.296	49.312	34.519	67.536
H7PA11QS2MGT	25.499	62.550	43.859	72.812
H7PB11QS2MGT	32.644	66.571	37.657	80.345
H7PA31QS2MGT	18.275	68.029	36.549	76.708
H7PB31QS2MGT	25.961	69.294	35.777	79.658
H7PA11QS3MGT	53.563	68.278	71.692	79.658
H7PB11QS3MGT	59.936	71.145	69.294	78.319
H7PA31QS3MGT	51.068	71.418	70.875	82.844
H7PB31QS3MGT	58.069	71.009	70.340	82.115

Table 5. Values of vertical static track stiffness at the beginning K_I and end of transition K_F (kN/mm) for all cases studied

These criteria were applied in two steps. Initially, the first two criteria were applied, and solutions were discarded if they resulted in either very high stiffness values, which generate elevated dynamic overloads, or very low stiffness values, which generate excessive rail deformations of the rail. To determine whether the stiffness values were high or low, they were compared to the values for designs using the same original ground material and the same materials as the simulated transition. Tab 6 shows the solutions remaining after this elimination process.

Materials of transition: Type1/Type 2	ORIGINAL GROUND QS1	ORIGINAL GROUND QS2	ORIGINAL GROUND QS3	ORIGINAL GROUND ROCK
QS2/QS3	PB31 PA11	PB31 PA11	PB11 PA31	PB11 PA31
QS2/MGT	PB31 PA11	PB31 PA11	PB11 PA31	PB11 PA31
QS3/MGT	PB31 PA11	PB31 PA11	PB31 PA31	PB31 PA31

Table 6. Transition types obtained after eliminating the transition types that with extreme stiffness values

The second step consisted of applying the third criterion, limiting the longitudinal variation value. From among the case studies, the cases selected were those with the smallest increase in the K value at the beginning and end of the technical blocks. In this way, the most appropriate solutions were obtained for each type of original ground material (QS2 and QS3). This approach yields certain design recommendations, which are described in the following section.

6. Proposed design recommendations

The analysis identified some problems related to the type of material. Based on these, the most relevant recommendations include:

- Excessive deformations were observed in the rail when material of type QS1 exists in the original ground (see Fig 11). These deformations reached 14mm under the rail when transitions QS2 and QS3 were used (Tab 7). In these cases, the deformations were large, as were the deformation values between adjacent sleepers. For this reason, it is appropriate to substitute the original ground material of type QS1 with another material or to treat the existing QS1 material in such a way as to obtain a modulus of elasticity corresponding to that of a material of at least type QS2.

In transitions from material of type QS2 to treated granular material, which occurs commonly in buried structures, the stiffness increased significantly at the beginning of the technical block when there was a relatively compressible material in the original ground. The solution with the smallest increase in the stiffness value was PA31. Fig 12 shows this value to be 16.4% which is too large. For this reason, transition-type QS2/MGT is not appropriate for use in the surface structure.

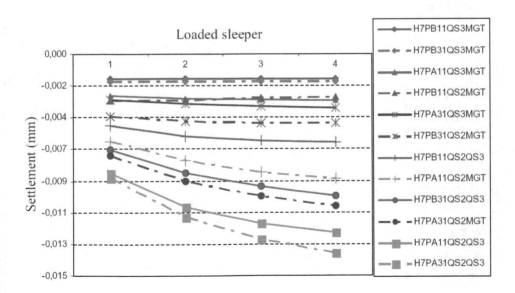

Fig. 11. Deflection of the head of loaded rails

	SLEEPER 5	SLEEPER 6	SLEEPER 7	SLEEPER 8
H7PA11QS2QS3				
LOADED SLEEPER 5	**-0.0083**	-0.0077	-0.0069	-0.0063
LOADED SLEEPER 6	-0.0104	**-0.0106**	-0.0100	-0.0091
LOADED SLEEPER 7	-0.0113	-0.0117	**-0.0118**	-0.0112
LOADED SLEEPER 8	-0.0117	-0.0120	-0.0123	**-0.0125**
H7PB11QS2QS3				
LOADED SLEEPER 5	**-0.0050**	-0.0045	-0.0038	-0.0032
LOADED SLEEPER 6	-0.0054	**-0.0057**	-0.0052	-0.0044
LOADED SLEEPER 7	-0.0051	-0.0057	**-0.0060**	-0.0055
LOADED SLEEPER 8	-0.0049	-0.0054	-0.0059	**-0.0062**
H7PA31QS2QS3				
LOADED SLEEPER 5	**-0.0086**	-0.0081	-0.0074	-0.0069
LOADED SLEEPER 6	-0.0111	**-0.0114**	-0.0109	-0.0101
LOADED SLEEPER 7	-0.0121	-0.0126	**-0.0129**	-0.0124
LOADED SLEEPER 8	-0.0127	-0.0131	-0.0136	**-0.0139**
H7PB31QS2QS3				
LOADED SLEEPER 5	**-0.0067**	-0.0062	-0.0055	-0.0049
LOADED SLEEPER 6	-0.0079	**-0.0083**	-0.0078	-0.0071
LOADED SLEEPER 7	-0.0083	-0.0089	**-0.0092**	-0.0087
LOADED SLEEPER 8	-0.0085	-0.0090	-0.0096	**-0.0099**
H7PA11QS2MGT				
LOADED SLEEPER 5	**-0.0061**	-0.0056	-0.0048	-0.0043
LOADED SLEEPER 6	-0.0072	**-0.0074**	-0.0069	-0.0062
LOADED SLEEPER 7	-0.0074	-0.0079	**-0.0082**	-0.0077
LOADED SLEEPER 8	-0.0076	-0.0080	-0.0084	**-0.0087**
H7PB11QS2MGT				
LOADED SLEEPER 5	**-0.0032**	-0.0027	-0.0020	-0.0015
LOADED SLEEPER 6	-0.0030	**-0.0032**	-0.0027	-0.0020
LOADED SLEEPER 7	-0.0026	-0.0029	**-0.0031**	-0.0026
LOADED SLEEPER 8	-0.0022	-0.0025	-0.0028	**-0.0030**
H7PA31QS2MGT				
LOADED SLEEPER 5	**-0.0071**	-0.0066	-0.0059	-0.0053
LOADED SLEEPER 6	-0.0086	**-0.0089**	-0.0084	-0.0077
LOADED SLEEPER 7	-0.0091	-0.0096	**-0.0099**	-0.0094
LOADED SLEEPER 8	-0.0094	-0.0098	-0.0103	**-0.0106**
H7PB31QS2MGT				
LOADED SLEEPER 5	**-0.0043**	-0.0038	-0.0031	-0.0025
LOADED SLEEPER 6	-0.0044	**-0.0047**	-0.0042	-0.0034
LOADED SLEEPER 7	-0.0040	-0.0045	**-0.0048**	-0.0043
LOADED SLEEPER 8	-0.0036	-0.0040	-0.0045	**-0.0048**
H7PA11QS3MGT				
LOADED SLEEPER 5	**-0.0029**	-0.0025	-0.0020	-0.0016
LOADED SLEEPER 6	-0.0028	**-0.0031**	-0.0027	-0.0022
LOADED SLEEPER 7	-0.0024	-0.0029	**-0.0032**	-0.0028
LOADED SLEEPER 8	-0.0021	-0.0025	-0.0029	**-0.0032**
H7PB11QS3MGT				
LOADED SLEEPER 5	**-0.0017**	-0.0014	-0.0010	-0.0007
LOADED SLEEPER 6	-0.0015	**-0.0017**	-0.0014	-0.0010
LOADED SLEEPER 7	-0.0011	-0.0015	**-0.0017**	-0.0014
LOADED SLEEPER 8	-0.0008	-0.0011	-0.0015	**-0.0017**
H7PA31QS3MGT				
LOADED SLEEPER 5	**-0.0032**	-0.0028	-0.0022	-0.0018
LOADED SLEEPER 6	-0.0031	**-0.0035**	-0.0031	-0.0025
LOADED SLEEPER 7	-0.0028	-0.0033	**-0.0036**	-0.0033
LOADED SLEEPER 8	-0.0025	-0.0029	-0.0034	**-0.0037**
H7PB31QS3MGT				
LOADED SLEEPER 5	**-0.0019**	-0.0016	-0.0012	-0.0009
LOADED SLEEPER 6	-0.0017	**-0.0019**	-0.0016	-0.0012
LOADED SLEEPER 7	-0.0013	-0.0017	**-0.0019**	-0.0016
LOADED SLEEPER 8	-0.0010	-0.0013	-0.0016	**-0.0019**

Table 7. Settlements of rail for the different load steps. Case study: original ground QS1

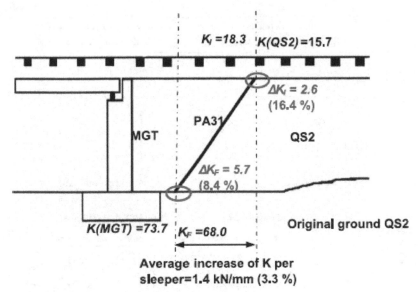

Fig. 12. Analysis of the stiffness variations assuming material transition from type QS2 to granulartreated material

- Near the abutment, the stiffness increased abruptly when using a granular material treated with cement as fill (MGT, 3% by weight). In fact, the stiffness value of the abutment could oscillate between 200KN/mm and 300KN/mm. The maximum of 87.8 KN/mm obtained for the case of treated granular material (Tab 4) is outside of that range. In this case, it is necessary to put a material with greater stiffness than MGT next to the abutment.

- A direct transition from QS2 or QS3 material to the structure is not recommended. At best, the stiffness is 38.43 KN/mm when the embankment material is QS2 and 74.027 KN/mm if it is QS3 (see Tab 4), as compared to200-300 KN/mm of vertical stiffness on the abutment.

- It is sometimes necessary to build the embankment adjacent to the abutment before building the structure. For example, this occurs when the original ground adjacent to the abutment is preloaded, and the load is removed immediately before building the transition. In that case, a slope of type PA (right column) is used instead of a slope of type PB (left column) in the transition from embankment material QS2 to QS3. In this research, a suitable value is obtained for the slope of the transition for each type (PA, PB). For that reason, it is necessary to distinguish between both slope types when considering whether a better approach would be to build the embankment before or after fabricating the structure.

In this way, based on these analyses, this study can make some recommendations about construction designs. Moreover, this study has proposed some ideas relating to the geometric designs of the different materials and has classified the designs according to the original ground and according to when the adjacent embankment is built, either before or after the construction of the structure (Fig 13).

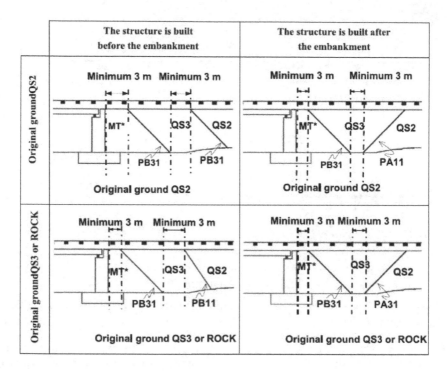

Fig. 13. Proposed design schemes

7. Conclusions

The designs currently used in embankment–structure transitions lead to zones that suffer significant deterioration. In addition, the various European railway systems have adopted many different design specifications when constructing these zones.

Increased railway speeds enhance the deterioration problems in the embankment–structure transitions, which has important implications for operating and maintenance costs, as well as for passenger safety and comfort.

From the numerical analysis carried out in this study, the following conclusions are drawn:

- The embankment must not be built on excessively compressible material, as shown for material of type QS1 on original ground. Such material must be replaced with another material or treated to obtain a modulus of elasticity corresponding to that of a material of at least type QS2.
- Transitions from material of type QS2 to material of type MGT should be carried out when there the original ground is of type QS3, or when the buried structure is treated.
- For each disposition type (PA, PB), there exists an optimal value for the slope. This value is not always the lowest one (3:1), as expected. The most suitable slope value (3:1 or 1:1) depends on the of original ground material, on the disposition type, and on the materials used in the transition.

- An abrupt increase in the stiffness takes place when a granular material treated with cement (MGT at 3% by weight) is used near the abutment. This problem remains unresolved, but future solutions should focus on improving the modulus of elasticity of the material without producing excessive stiffness increases at the extremes of the transition from the material of type QS3 to the treated material.

The application of all of those conclusions leads to a succession of recommendations for the most suitable building designs. When developing such designs, those factors which have an influence on the transition behavior (original ground, transition materials, and slope type) must be considered.

8. References

Amielin, S. (1974). La vía y el servicio de vía. (Spanish translation from the original in Russian, by Fundación de los Ferrocarriles Españoles).

Comité D-117 (ORE). (1983). Adaptation optimale de la voie classique au trafic de l'avenir.Rapport no. 27. Comportement des structures d'assise de la voiesous charges répétées. Office de Recherches et d'Essais del'Union Internationale des Chemins de Fer.

Esveld, C. (2001). Modern railway track. The Netherlands: MRTProductions.

European comunity. (2005). Sustained Performance of Railway Tracks-SUPERTRACK Project Numerical simulation of train-track dynamics.

European Rail Research Institute, ERRI. (1999). Code D-230.1/RP3, Bridge ends embankment. Structure transitions.

Gallego, I. (2006). Heterogeneidad resistente de las vías de Alta Velocidad: Transición terraplén-estructura.6. Ciudad Real: Doctoral Thesis, University of Castilla-La Mancha.

Gallego, I., & López, A. (2009). Numerical simulation of embankment–structure transition design. Journal of Rail and Rapid Transit, 223, 331-342.

Gallego, I., López, A., Ubalde, L., & Texeira, P. (2005). Track deterioration in high-speed railways: the influence of longitudinal variatioical stiffrness in the embankment-structure transitions. Congress on Railway Engineering. London.

Hettle, A. (1986). Modelluntersuchungen der Gleissetzungen am Ubergang Brücke (Studies in a reduced scale model on the track sinking/downfall/collapse in a section of a bridge/embankment transition). ETR (Spanish translation from the original in German) .

Hunt, H. (1997). Settlement of railway track near bridge abutment. Proc. Instn. Civil Engrs. Transp, (págs. 68-73).

López, A. (1977). Análisis de la deformabilidad vertical de una vía férrea mediante el método de elementos finitos. AIT (15).

Lopez, A. (1983). La heterogeneidad resistente de una vía y su incidencia en la evolución de la nivelación longitudinal: una aproximación al problema. Revista de Obras Públicas, , 719-735.

López, A. (2001). La rigidez vertical de la vía y el deterioro de las líneas de alta velocidad. Revista de Obras Públicas , 222-232.

Ministerio de fomento. (1999). Recomendaciones para el proyecto de plataformas ferroviarias. Madrid: Servicio de publicaciones del Ministerio de Fomento.

Mira, P., Férnández, J., Pastor, M., Nasarre, J., & Carrillo, J. (2000). Aplicaciones del método de elementos finitos a la ingeniería ferroviaria. Revista de Ingeniería Civi (118), 71-82.

Oliver, X., & Arlet, C. (2000). *Mecánica de medios continuos para ingenieros*. Barcelona: Ediciones UPC.

Profillidis, V. (1983). *La voie et sa fondation modelisation mathematique*. París: Doctoral Thesis, École Nationale des Ponts et Chaussées.

Prud`Homme, M. A. (1970). La voie. *Revue Générales des Chemins de Fer* .

Sahu, J., Rao, & Yudhbir. (1999). Parametric study of resilient response of tracks with a sub-ballast layer. *Canadian Geotechnical Journal , 36*, 1137-1150.

Sauvage, G., & Larible, G. (1982). La modélisation par éléments finis des couches d`assise de la voie ferrée. *Revue Générale des Chemins de Fer* , 475-484.

Teixeira, P. F. (2003). *Contribución a la reducción de los costes de mantenimiento de vías de alta velocidad mediante la optimización de su rigidez vertical*. Doctoral Thesis, Polytechnic University of Cataluña.

## 2

Evolutionary Algorithms in Embedded Intelligent Devices Using Satellite Navigation for Railway Transport

Anatoly Levchenkov, Mikhail Gorobetz and Andrew Mor-Yaroslavtsev

Riga Technical University
Latvia

1. Introduction

Nowadays, the most widely spread type of a computer is an embedded system. The embedded systems consist of the following hardware (i.e. nano-electronic components) – programmable microcontrollers or microprocessors; transmitters, including the receivers of global positioning information systems, which demonstrate the state and measuring parameters of a controlled object, and which relay that to the programmable microcontroller; actuators, which receive a signal from the programmed microcontroller and relay it to an antenna, a display or an electro-drive device; and communication devices, including wireless communication with other devices and the software with algorithms of artificial intelligence (Russel, Norvig, 2006).

Railway traffic flow is limited by safety criteria. Therefore, routing and scheduling task is actual for a railway transport. As well an optimal braking control and safety of braking process are very important (Luo, Zeng, 2009). The analysis of human behavior and simulation of train's braking are investigated (Hasegawa et al., 2009). An intelligent transport control system gives a possibility to make traffic control safer and more cost-effective (Gorobetz, 2008). It may find an optimal solution to a conflict faster than a human as a decision support system (Levchenkov et al., 2009). In addition in case of emergency it may prevent crashes and accidents without human intervention.

Authors propose the intelligent braking control device, which warns the driver about the necessity of starting the working braking, taking into account the signal of the traffic light. If working braking has not been started the controller activates emergency brakes with a purpose to stop before the beginning of the next block-section if it is possible, or to choose a free way with enough free distance to stop without a crash. The primary task of the diagnostics device is to separate dangerous situations in braking system by critical values of sensor measurements from the regular states of the system, to detect and to warn about changes in the system and to prevent emergencies immediately. The system allows stopping the train timely before the problem has occurred.

2. Problem formulation

Railway safety is an actual and important task. Nowadays a human factor is the main reason for 74% of railway accidents and crashes. This problem is actual as in Latvia as all over the world. Various crashes like in Riga (Latvia) in February 2005, in Ventspils (Latvia) in December 2008, in Aegvidu (Estonia) in December 2010, in Brussels (Belgium) in February 2010, in Magdeburg (Germany) in January 2011 prove the necessity of finding the problem solution.

The main reason of railway accidents is a human factor, when

- train driver does not stop a train on the restrictive signal
- incorrect decisions were made by signalmen (station-on-duty) in extraordinary situations

Therefore, auxiliary embedded electronic devices are needed

- to help a train driver and station signalmen to make the best decision faster
- to prevent accidents if a human does not react

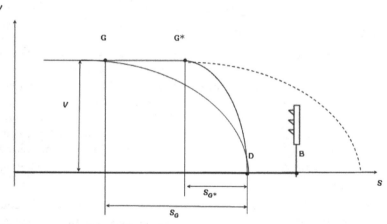

Fig. 1. Graphical representation of the task.

The primary task is to stop the train before the restrictive signal. For this task a warning point G should be detected, where regular braking still may be applied (Fig. 1.). The point G* is a marginal point and also must be defined, where only emergency braking may be applied to stop the train before the signal. Using of regular braking in G* follows the passing of the restrictive signal.

The purpose of the research is to develop a prototype of a new control command and a signalling track-side and on-board locomotive's devices to improve train movement safety.

The following functions of the on-board locomotive's embedded device are required:

- receiving location and other necessary data from satellite navigation system
- reception of the data from the server about railway infrastructure (tracks, points) and control points (signals, section points) in the location of the train;
- selecting the necessary signal on the way;
- establishing the wireless connection with the selected signal (Fig. 2.)

- receiving the data from a track-side signal
- checking if the train has reached the route control point;
- detecting the acceleration of the train
- detecting the regular braking distance;
- detecting the emergency braking distance;
- detecting the speed limitation;
- detecting the starting point of regular braking;
- starting emergency braking process
- a module for checking the reaching of the starting point of regular braking;
- warning about the necessity to start regular braking;

The following functions of the track-side embedded device are required:

- establishing the wireless connection with the train
- transmitting signal value to the locomotive's device
- detection of the safest state of the station points using evolutionary algorithms to avoid the collision (Fig. 3.).

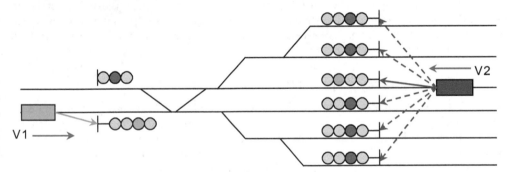

Fig. 2. Selection of the necessary signal on the way.

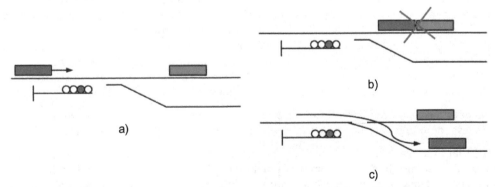

Fig. 3. Detection of the safest state of the station points to avoid the collision.
a) Initial state, b) Collision is possible, c) Collision is avoided.

There are many embedded transport control systems on the market which are designed to provide safety for a vehicle, its passengers or cargo, and other traffic participants.

In the commercial railway transport segment an example of such a safety system is KLUB-U, currently used on Russian Railways. It is installed in the locomotives and by interacting with existing signaling systems and its own modules provides information about the train's and its closest neighbors' coordinates, diagnostics of the brakes, current railway segment profile and maximum allowed speed, and controls the vigilance of the locomotive driver. Still, despite the wide array of features, it lacks automation and many decisions require manual operation.

A significant component of the whole safety system is the circuits, engines and brakes diagnostics complex. While the most complete diagnostics can be performed only in the technical service environment, most failures can be detected during its operation using circuit integrity indicators and different sensors designed to uncover electrical mechanical damage.

All kinds of damages which could lead to failures can be combined into distinctive value sets, thus recognizing them in the stream of incoming data allows early identification of problems in the engine.

Artificial immune systems (AIS) were mentioned in some papers in mid 1980s but became a subject in its own right in 1994 in the papers on negative selection (Forrest et al., 1994, Kephart, 1994). Currently the systems are actively explored for possible use cases. For example, there are studies on a real-valued negative selection algorithm for an aircraft fault detection (Dasgupta et al., 2004).

3. Structure of proposed system for railway safety tasks

The chapter will demonstrate some issues of design and modelling of a part of a modern embedded system for a rail transport (Fig. 4.). This embedded system is intended for managing the rail transport's electrical drive and the traffic lights, and it consists of the microcontrollers, the developed software and the information system, the wireless communication possibilities and the global positioning system.

Fig. 4 illustrates a complete scheme of a structure of the rail transport's embedded system. In the figure, the brown colour shows the train's embedded devices; the green colour stands for the devices of the traffic lights' embedded system; the purple colour – the devices of the crossing's embedded system; the yellow colour – coordination embedded system devices; the light blue colour – software which provides operation of the network of the wireless embedded system, operation of the communication network of the wireless devices and which is responsible for making operational decisions.

The device receives the signal from the defined traffic light and defines its position using a wireless communication network and has data storage with route control points as well.

Artificial immune systems use evolutionary data processing paradigm based on biological immune systems. It differs from computational immunology which models biological immune systems. Immune algorithms are mainly used to solve anomaly recognition, data collection and analysis tasks. From the computational point of view the most interesting features of the immune systems are self-learning, diversity maintenance and memory.

The problem is represented as an antigen and a solution candidates as antibodies which are randomly generated from the library of available solutions or genes. The evaluation of affinity or degree of binding between the antigen and the antibody is similar to complementarity level in biological IS and it defines the fate of each individual antibody as well as the termination of the whole algorithm. Individual antibodies are replaced, cloned and hypermutated until a satisfactory level of affinity is reached. Partial replacement of the solutions' population with fresh randomly generated candidates maintains diversity which allows solving a wider set of problems. The probability of cloning or hypermutating a candidate depends on its affinity.

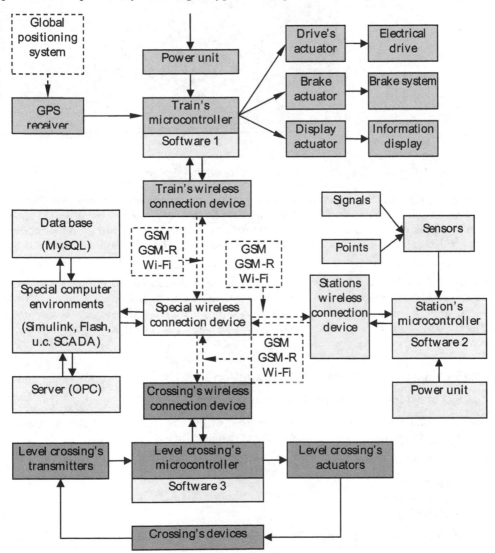

Fig. 4. Wireless network structure's scheme of embedded devices for a rail transport

In the rolling stock safety system (Fig. 5) (Mor-Yaroslavtsev, Levchenkov, 2011), the invading object I is picked up by sensors S and the data is transmitted to the nearest cell tower *CT*, which relays it to the control center *CC* and the nearest locomotives wireless modems *M*. Through the same modem the locomotive *L* receives data about the closest neighbors' rolling stock position and status, railway segment profile and maximum allowed speed.

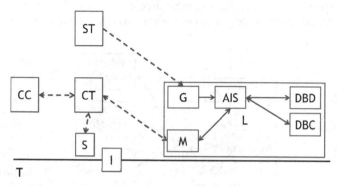

Fig. 5. The intelligent rolling stock safety system functional design

L also hosts: a positioning receiver *G* which receives data from a positioning satellite *ST*; data analysis module *AIS* which communicates to the immune detector database *DBD* and control cell database *DBC*. Depending on the results of control cell maturation the module makes a decision and executes it by sending a control signal or displaying an alert to the driver.

Analogous to the hybrid IDS (Powers, He, 2009) the most feasible way to implement such a system would be through the two phases of anomaly detection and determination of their type to draw a conclusion. In this case the incoming data from the sensors is the set of antigens. The data includes but is not limited to speed, acceleration, voltage, rotation, temperature, and presence of smoke.

4. Mathematical models for problem solution

4.1 Model of differential positioning system

Differential satellite navigation systems are used to increase precision of the positioning systems that is very significant for safety-critical systems, such as transport.

Differential satellite navigation systems (Fig. 6.) contain the following object types:

-   S – satellite;
-   M – base station;
-   R – receiver.

Each satellite S is described by the following parameters:

-   α – slope of the satellite's orbit to the equator plane;
-   Ω – slope of up-going node of the satellite's orbit to the Greenwich meridian;
-   ω – perigee angle from up-going node
-   t – time of crossing perigee or up-going node of the satellite's orbit;

- e – eccentricity of the satellite's orbit:

$$e = \sqrt{1 - (b / a)^2} \, ,$$ (1)

where a and b are half-axes of the ellipsoid orbit;

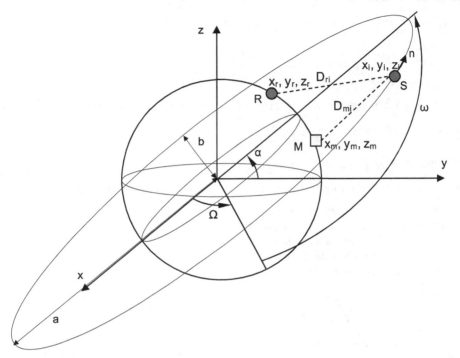

Fig. 6. Differential satellite navigation system elements

- n – angular velocity of the satellite,

$$n = \sqrt{\mu / a^3} \, ,$$ (2)

where μ – Earth gravimetric constant.

- x, y, z – coordinates of the satellite

Base station M parameters:

- x_m, y_m, z_m – coordinates of the base station
- D_{mi} – known distance between i-th satellite and base station m:

$$D_{mi} = \sqrt{(x_i - x_m)^2 + (y_i - y_m)^2 + (z_i - z_m)^2}$$ (3)

- ρ_{mi}, $\Delta\rho_{mi}$ - distance measurement result and necessary correction between measured and real distance:

$$\Delta\rho_{mi} = \rho_{mi} - D_{mi} = \varepsilon_{m,sat} + \varepsilon_{m,con} + \varepsilon_{m,rec} + c \cdot \delta t_m , \tag{4}$$

where $\varepsilon_{m,sat}$ - satellite apparatus error, satellite clock error, $\varepsilon_{m,con}$ - control error, incorrect ephemerid forecast, $\varepsilon_{m,rec}$ - receiver's error, ionosphere, troposphere and other noises, δt_m - base station clock deviation from satellite clock, c - light speed.

- ρ_{ri} - corrected distance measurement between recipient and satellite:

$$\rho_{ri} = D_{ri} + \varepsilon_{r,sat} + \varepsilon_{r,con} + \varepsilon_{r,rec} + c \cdot \delta t_m - \Delta\rho_{mi} = D_{ri} + \varepsilon_r + c \cdot \delta t_{mr} =$$
$$\sqrt{(x_i - x_r)^2 + (y_i - y_r)^2 + (z_i - z_r)^2} + \varepsilon_r + c \cdot \delta t_{mr} \tag{5}$$

where ε_r – receiver's result segment error, δt_{mr} – combined clock deviation,

x_r, y_r, z_r – coordinates of the receiver.

4.2 Model of railway station

The model of the station may be described with the following sets of objects:

- $V = \{v1, v2, ..., vn\}$ – set of trains;
- $M = \{m1, m2, ..., mn\}$ – set of train goals;
- $L = \{l1, l2, ..., lk\}$ – set of signals;
- $P = \{p1, p2, ..., pq\}$ – set of points;
- $R = \{r1, r2, ..., rw\}$ – set of routes.

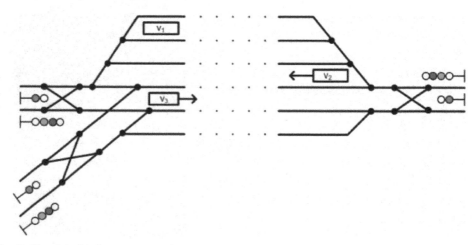

Fig. 7. Graphical interpretation of station model

4.3 Model of braking of rolling stock

The braking way consists of preparation and real segments:

$$S_T = S_P + S_D ; \tag{6}$$

where S_T is the braking distance, S_p – distance of moving during the preparation of braking system, S_D – real braking distance

The braking power of the train should be defined taking into account a real force of the braking chock influencing the train wheels. A real friction factor depends on the braking chock material.

The following factor characterises the braking chock made of cast iron:

$$\varphi_K = 0,6 \frac{16K + 100}{80K + 100} \cdot \frac{v + 100}{5v + 100};$$

(7)

The cast iron braking chocks containing phosphorus of 1,0-1,4% are characterised with the factor

The mentioned above factor for braking chocks of composite materials can be defined with the following expression :

$$\varphi_K = 0,44 \frac{K + 20}{4K + 100} \cdot \frac{v + 100}{2v + 100};$$

(8)

The calculations of braking force of the chocks also depend on a type of chocks.

For the standard cast iron braking chocks:

$$K_p = 2,22K \frac{16K + 100}{80K + 100};$$

(9)

The cast iron- phosphorus braking chocks:

$$K_p = 1,85K \frac{16K + 100}{52K + 100};$$

(10)

For the braking chocks of composite materials:

$$K_p = 1,22K \frac{K + 20}{4K + 20};$$

(11)

The total braking factor:

$$\vartheta_{PO} = \frac{\sum K_P}{Q + P_u};$$

(12)

The main force resistive to the motion in idle running

$$W_{Ox} = W_{O1} + N \cdot W_{Oc};$$

(13)

where N - quantity of the carriages.

Locomotive without the train:

$$W_{0l} = 24 + 0,11v + 0,0035v^2 ; \tag{14}$$

Cargo carriages:

$$w_{0c} = 7 + \frac{a + v + 0,025v^2}{q_0} ; \tag{15}$$

Passenger trains:

$$w_{0c} = 12 + 0,12v + 0,002v^2 . \tag{16}$$

As within the time interval Δt the braking force and the opposite self-resistive force ω_{Ox} to the motion of the train are assumed as constant values then the increasing of the speed can be calculated according to :

$$\Delta v = \frac{\xi(b_T + \omega_{Ox} + i_c)\Delta t}{3600} . \tag{17}$$

The speed of the braking force distribution is a braking wave: $v_t = \frac{L}{t_t}$; where: L is the length of the train; t_t – time from the moment when the driver turns the handle of the hoist till the pressure appears (?) in the braking cylinders; air wave : $v_v - 20\sqrt{T}$; where: $T = 273 + t^\circ C$ – absolute temperature of gas.

The preparation braking distance:

$$S_P = 0,278v_0 t_p = \frac{v_0 t_n}{3,6} . \tag{18}$$

Real braking distance:

$$S_D = \sum \frac{500(v_N^2 - v_K^2)}{\zeta(w_{ox} + b_m + i_C)} ; \tag{19}$$

Thus the total braking way:

$$S_T = \frac{v_0 t_n}{3,6} + \sum \frac{500(v_N^2 - v_K^2)}{\zeta(w_{ox} + b_m + i_C)} \tag{20}$$

4.4 Model of railway infrastructure and command and control system

Rail ways can be represented as a graph $R = \{C,S\}$, where rails are divided into sections S, and each section $s \in S$ is connected with each other by two connectors $s =< c_i, c_j >$.

Each section $s \in S$ has a constant length l_s , a curve a_s , and a speed limit v^*_s .

Each point $p \in C$ has a connecting set W of three or more sections and a set of possible states of point D_p, where $d_p^n = <s_i, s_j>$ means opened in both directions from s_i to s_j and from s_j to s_i is following for different point types:

single point: $D_p = \{<s_i, s_j>, <s_i, s_k>\}$;

dual point: $D_p = \{<s_i, s_j>, <s_i, s_k>, <s_i, s_m>\}$;

cross point: $D_p = \{<s_i, s_j>, <s_i, s_k>, <s_m, s_j>, <s_m, s_k>\}$.

Each state of point $d_p \in D_p$ has a speed limit $v^*_{d_p}$; maximal each point's $d_p \in D_p$ switching time: t_{d_p} .

Railway signal G is an object with fixed coordinates x_0, y_0 connected to the fixed position on the track.

Each signal $g \in G$ has the following states of signals $L_g \subseteq \{R, Y, YG, G, V, W\}$, where "R" – red, and rolling stock must stop before the signal; "Y" – yellow, can move and be ready to stop, the next signal is red; "YG" – yellow and green, next two sections are free; "G" – green, "V" – violet, "W" – moonlight white.

Each signal sets up speed limits for the next block-section: v_{def}- maximal predefined speed on the section, v_0- 0 kmh, stop; v_1- < 50 kmh, movement on turnouts 1/9 and 1/11 types; v_2- < 80 kmh for movement on turnout 1/18 type; v_3 - < 120kmh for movement on turnout 1/22 type.

4.5 Assessment functions

Multi-criteria target function for braking:

$$
\begin{cases}
F^{br}(DL, CL, EL, \Delta) \to min \\
DL = \Delta S \to 0 \\
CL = \varepsilon(t) \to \varepsilon * \\
EL = \dfrac{d\varepsilon}{dt} = const \\
\Delta \to \{\bar{0}, C_1, C_2, ...\} \\
\Delta = |\Phi(t) - E(t)| \\
E(t) = \{\vartheta_p, S_B, \varepsilon_i, t_m, \tau, I_{dcp}, Q_v\} \\
\Phi(t) = \{\vartheta^f, S_B^f, \varepsilon_i^f, t_m^f, \tau^f, I_{dcp}^f, Q_v^f\}
\end{cases}
\tag{21}
$$

- ΔS – distance between the closed section and rolling stock – danger level criteria (DL)
- $\varepsilon(t)$ – deceleration of rolling stock
- $\varepsilon *$ – optimal deceleration for passengers – comfort level criteria (CL)
- $d\varepsilon / dt$ – changes of deceleration and braking torque – optimal energy consumption criteria (EL)

- Fbr – function for braking process optimization
- Δ – difference/similarity vector, which defines the difference between the estimated and the actual values.
- $\bar{O}$ – Zero vector, which means a compliance with the normal situation, where the estimates and the actual value of the difference is zero;
- C_i – situation of danger classes, according to differences between estimated and the actual values;
- $E(t)$ – vector of estimated values;
- $\Phi(t)$ – vector of actual values;
- f – index, which represents the actual values.

The routing task for accident prevention consists of a generation of a new route and schedule for rolling stocks V moving on points P.

The target function for scheduling and routing is to arrange points for each train to reach a destination and assigning of time moments t to each train and each point.

- Train's schedule: $\sigma_v : P \rightarrow \{t_{v1}, t_{v2}, ..., t_{vs}\} \subset \Re$
- Point's schedule: $\sigma_{p^1} : V \rightarrow \{t_{p1}, t_{p2}, ..., t_{pm}\} \subset \Re$

The target function for an optimal point state on the station is the following:

$$T_\Sigma = f(t_1, t_2, ..., t_n; \quad x_{11}, x_{12}, ..., x_{1q}; \quad ... \quad x_{n1}, x_{n2}, ..., x_{nq}) \rightarrow \min \qquad (22)$$

t_i – the i-th time moment of switching points
x_{ij} – state of the j-th point in the i-th time moment

5. Evolutionary algorithms for problem solution

5.1 Fitness function for genetic algorithm

Step 0. Initialization

$T_{i\,sum} = 0$ – for each i-th train summary time

$G_i = 0$, for each i-th train goal achievement

$i = 1$ – selected train number

Step 1. Moving the time calculation of the i-th train on the j-th railway section, $t_j^i = S_j / v_i$ where

S_j – length of the j-th section,

v_i – current i-th train speed

Step 2. Check if the j-th section ends with signal.

Step 2.1. Check the occupancy of all tracks to the next signal.

Step 2.2. If any of sections in Step 2.1 is busy and the train is moving, then recalculate time with braking conditions and Goto 4.

$$S_b^i = \frac{-v_i^2}{2\varepsilon}; \quad S_r = S_j - S_b; \quad t_r = \frac{S_r}{v_i}; \quad t_b = \frac{v}{\varepsilon}; \quad t_j^i = t_r + t_b; \quad v_i = 0$$

Step 2.3. If train is already stopped then check If all sections in Step 2.1 are free, then recalculate time with acceleration conditions

Step 2.4. $j = j + 1$, select next section of the route

else if the j-th section is not a signal then $j = j + 1$, select the next section of the route

Step 3. Check if the train reached the goal, then $Gi = 1$

Step 3.1. Check if all trains reached the goal then $T_\Sigma = \max(T_{SUM}^1, T_{SUM}^2, ..., T_{SUM}^n)$

Step 3.2. Function STOPS

Step 4. Update summary time of i-th train $T_{SUM}^i = T_{SUM}^i + t_j^i$

Step 5. Check if the summary time is less than the next point switching time $T_{SUM}^i \le t_p$?

Step 5.1. If $T_{SUM}^i \ge t_p$ then $T_{SUM}^k = \min(T_{SUM}^1, T_{SUM}^2, ..., T_{SUM}^n)$

Step 5.2 If $k = i$ then Goto 6, else $i = k$, Goto 1

Step 6. Check occupancy of tracks in switching moment

Step 6.1. Recalculate position of the train $S_{beg}^i = v_i \cdot (T_{SUM}^i - t_j^i - t_p)$

Step 6.2. Find "tail" of the train $S_{end}^i = S_{beg}^i - L_i$

Step 6.3. Check the occupancy of all tracks.

If switching points are busy, then $T_\Sigma = \infty$, algorithm ENDS; else Goto 1

5.2 Genetic algorithm

A genetic algorithm for the task solution may be described with the following steps.

1 step: Initialize random set of possible solutions: $S^{(0)} = \{s_1^{(0)}, s_2^{(0)}, ..., s_{s_{max}}^{(0)}\}$;

2 step: Evaluate each solution with a target function: $V^S = \{F(s_1), F(s_2), ..., F(s_p)\}$;

3 step: Arrange solutions by evaluation: $\overline{S} = \{\overline{s}_1, \overline{s}_2, ..., \overline{s}_p\}$, $F(\overline{s}_1) = opt(V^S)$;

4 step: Duplicate the best solutions in the elite set: $S_E \subset \overline{S}$;

5 step: **Selection.** Select from the set of solution pairs according to the defined selection: $S_C = \overline{S}$;

6 step: **Crossover:** Generate a new population from the set of the solution pair according to the defined crossover algorithm:

$$\overline{s}_i \Pi \overline{s}_j \rightarrow s'_i = s_{ij}; s'_j = s_{ji}, \quad i, j = \overline{1, p};$$ (23)

7 step: **Mutation:** Random change of one of solution parameter that helps to find a global optimum of the function:

$$x_j^{s'_i} = x_j^{s'_i} + 1, \quad s'_i \in S', \quad j = rand(\overline{1, k}), \quad i = rand(\overline{1, p});$$ (24)

8 step: Evaluate the new population using the target function:

$$V^{S'} = \{F(s'_1), F(s'_2), ..., F(s'_p)\};$$ (25)

9 step: Arrange the new population by the evaluation values:

$$\overline{S}' = \{\overline{s}'_1, \overline{s}'_2, ..., \overline{s}'_p\}, \quad F(\overline{s}'_1) = opt(V^{S'});$$ (26)

10 step: Add the new population of solution to the elite set: $S = (S_E + \overline{S}');$

11 step: Delete the last solutions from the population S if its size exceeds predefined population size p: $S = S / \{s_{p+1}, s_{p+2}, ...\};$

12 step: Algorithm stops by time, generation, convergence or by another predefined criteria. If stop criteria is false then repeat the algorithm from step 4. If true then the result of the algorithm is solution s_1.

5.3 Algorithm for an artificial immune system

5.3.1 Shape-space concept

AIS are modeled after biological IS and carry the terms of antigens and antibodies. They can be modeled using the shape-space concept (see Fig. 8.) (Musilek et al., 2009). The shape-space S allows defining antigens, receptors and their interactions in a quantitative way.

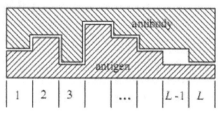

Fig. 8. A shape-space model of an antigen and an antibody.

Like chromosomes in the in the evolutionary algorithms, depending on a problem being solved it also could be a set of integers or binary numbers – $m \in Z^L$ or $m \in \{0,1\}^L$.

The affinity of an antigen–antibody pair is related to their $\in$ distance in the shape-space S and can be estimated using any distance measure between the two attribute strings. The distance between an antigen, Ag, and an antibody, Ab, can be defined, for example, using a general class of Minkowski distance measures:

$$D_M(Ag, Ab) = \sqrt[p]{\sum_{i=1}^{L} |Ag_i - Ab_i|^p}$$

By varying the value of the parameter p a suitable measure of distance can be obtained.

5.3.2 The negative selection algorithm

Negative selection is the paradigm describing the evolution of the T-lymphocytes where they are randomly generated and learn to recognize all except the self structures, specific to the host. Negative selection algorithms need training samples only from one class (self, normal), thus, they are especially suited for the tasks such as novelty, anomaly or change detection including those in engines and other devices.

The key advantage of anomaly detection systems is their ability to detect novel attack patterns for which no signature exists, while their most notable disadvantage is a larger false positive rate.

The algorithm:

Step 1. Define a set S which needs to be monitored and the set P of the know self $m = \langle m_1, m_2, ..., m_L \rangle \in S^L$ elements in a feature space U. The set U corresponds to all the possible system states, P – normal states and S – the current state which changes in time.

Step 2. Generate a set of candidate detectors $C = \{c_1, c_2, ..., c_n\}$.

Step 3. Compare each candidate c_i to the set of known good elements P.

Step 4. If a match occurs, discard the individual c_i, otherwise store it in the mature detector set D. Or, to maximize the nonself space coverage with a minimum number of detectors, move the matched candidate away from the closest element p_j, then store it in D.

Step 5. Monitor S for changes by continually matching it against the detectors in D. If any detector matches, the change which has occurred most likely is dangerous, as D is designed not to match any normal system state.

This algorithm produces a set of the detectors capable to recognize non-self patterns. The action following the recognition varies according to the problem under consideration. In the case of transport safety control system it could be an alarm or issue of an immediate stop signal depending on the detected fault.

The detectors and the caught fault conditions are stored in an immune memory for further processing and to provide further information about the consequences of the attack and possible future actions instead of simply reporting the incidents.

6. Computer experiments

6.1 Computer experiment of genetic algorithm

The task of the experiment is to minimize idle time of trains on the station and to minimize the risk of their collision.

The station (Fig. 9.) with 4 points p_1, p_2, p_3, p_4 is given and two trains V1 and V2 are approaching. Railway tracks of the station are split into the sections, where start and end of each section is a point or a signal. The length of the trains $L_{v1} = 500$ m and $L_{v2} = 300$ m is given.

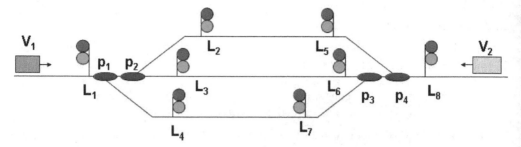

Fig. 9. Structure of the station for the computer experiment

Generation 1	Generation 12
0) 0011000010010010000101000100	**0)** 1000110001001101000001000100
f(x) = 610.47947316824; T1 = 379, T2 = 564 Points states: 0 1 0 0 | 0 1 0 0	f(x) = 270; T1 = 96, T2 = 406 Points states: 0 1 0 0 | 0 1 0 0
1) 0011000010010010000101000010	**1)** 1000110000000101001001000100
f(x) = 610.47947316824; T1 = 379, T2 = 564 Points states: 0 1 0 0 | 0 0 1 0	f(x) = 270; T1 = 94, T2 = 160 Points states: 0 1 0 0 | 0 1 0 0
2) 0110010000001101000101000100	**2)** 1000111000000101000001001111
f(x) = 639.47947316824; T1 = 781, T2 = 408 Points states: 0 1 0 0 | 0 1 0 0	f(x) = 270; T1 = 109, T2 = 156 Points states: 0 1 0 0 | 1 1 1 1
3) 0011101001011101000101001111	**3)** 1000110010000101010101000110
f(x) = 686.47947316824; T1 = 455, T2 = 908 Points states: 0 1 0 0 | 1 1 1 1	f(x) = 270; T1 = 98, T2 = 166 Points states: 0 1 0 0 | 0 1 1 0
4) 1100111101111000110101000100	**4)** 1000110000001010101001000100
f(x) = 850.47947316824; T1 = 619, T2 = 775 Points states: 0 1 0 0 | 0 1 0 0	f(x) = 270; T1 = 94, T2 = 332 Points states: 0 1 0 0 | 0 1 0 0
5) 0101110101111010100101000100	**5)** 0001000010011010000101000110
f(x) = 960.47947316824; T1 = 729, T2 = 830 Points states: 0 1 0 0 | 0 1 0 0	f(x) = 270; T1 = 129, T2 = 814 Points states: 0 1 0 0 | 0 1 1 0
6) 1101101011000001110001011101	**6)** 1000110010000101010101000010
f(x) = 100000; T1 = 709, T2 = 55 Points states: 0 1 0 1 | 1 1 0 1	f(x) = 270; T1 = 98, T2 = 166 Points states: 0 1 0 0 | 0 0 1 0
7) 0110110101000011011010100000	**7)** 1000110001001101001001000100
f(x) = 100000; T1 = 854, T2 = 105 Points states: 1 0 1 0 | 0 0 0 0	f(x) = 270; T1 = 96, T2 = 410 Points states: 0 1 0 0 | 0 1 0 0
8) 1100110000101010110001100101	**8)** 1000110111000101010001000100
f(x) = 100000; T1 = 594, T2 = 336 Points states: 0 1 1 0 | 0 1 0 1	f(x) = 270; T1 = 107, T2 = 164 Points states: 0 1 0 0 | 0 1 0 0
a)	b)

Fig. 10. Results of Genetic Algorithm - a) first iteration, b) last iteration

The fitness function and the algorithm are realised in the program and the following parameters for genetic algorithm are used:

- Crossover rate - 0.8;
- Mutation rate - 0.01;
- Population size - 50
- Random parent selection
- Single point crossover

Table 1. shows the dynamics of genetic algorithm. The algorithm is performed in 2 seconds and the algorithm converges completely in the 12th generation, where an average value of the population is equal with the best value.

Generation	Average	Best
1	88087.15753678	610.47947316824
2	82128.38630517	413.47947316824
3	70197.754252487	270
4	50320.950968194	270
5	32391.38814628	270
6	26405.504351969	270
7	14391.055583579	270
8	2366.4284573351	270
9	2332.525484097	270
10	329.57671570692	270
11	307.08794731682	270
12	270	270

Table 1. Average and best values of fitness function on each generation

6.2 Computer experiment for an artificial immune system

6.2.1 Collecting the location data

One data set for the experiment was taken from the two PLCs in the field attached to a vehicle and a level crossing. The data collection scheme is presented in Fig. 11.

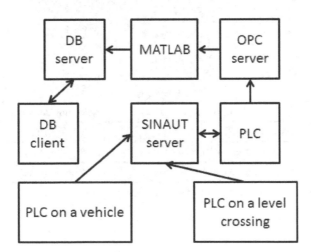

Fig. 11. The location data collection scheme.

The communication between the PLCs is facilitated by GPRS modules and a server running on a PC. Through the chain of software tools the data is piped from the PLCs to DB tables.

The data was collected into two tables for records related to a vehicle and a level crossing. The report (Fig. 4) contains this data cross-matched using the date and time — as all the data was simultaneously recorded with discrete steps of 1 s, the matches are 1:1.

This provides a set of data for test runs of the algorithms.

6.2.2 The real-value negative selection algorithm

The RNS detector generation starts with a population of candidate detectors, which are then matured through an iterative process. In particular, the center of each detector is chosen at random and the radius is a variable parameter which determines the size of the detector in m-dimensional space. The basic algorithmic steps of the generation algorithm are given in 5.3.2.

The whole detector generation process terminates when a set of mature (minimum overlapping) detectors are evolved which can provide significant coverage of the non-self space.

A detector is defined as $d = (c, r_d)$, where $c = (c_1, c_2, ..., c_m)$ is an m-dimensional point that corresponds to the center of a hypersphere with r_d as its radius. The following parameters are used (Fig. 12):

- r_s: threshold variation of a self point;
- a: variable movement of a detector away from a self sample or existing detectors;
- ξ: maximum allowable overlap among the detectors, allowing some overlap can reduce holes in the non-self coverage.

Settings:

- Maximum self-element variation: 0.2
- Maximum detector overlap: 0.1
- Dimensions (sensors): 4
- Maximum detector population: 10
- Number of tests: 20
- Next generation after 5 tests
- Number of top detectors to clone: 2

Problem space

Dimension	Minimum	Maximum
1. Train speed, km/h	0	100
2. Crossing car speed, km/h	0	150
3. Distance from the train to the rendezvous, m	-100	1000
4. Distance from the car to the rendezvous, m	-10	1000

Self elements

- [0, 5, 1000, 0]
- [50, 30, 900, 10]
- [20, 30, 600, 2]

Fig. 12. A screenshot from the computer program running a real-valued negative selection algorithm showing the initial settings for training the detector set.

During the straightforward detection process the matured detectors are continually compared to new test data samples. The distance D between a sample pattern $p = (c_p, r_s)$ and a detector $d = (c_d, r_d)$ is computed in the same way as in the detector generation phase. If $D < (r_s + r_d)$ then the detector d gets activated indicating possible fault.

#	Centre	Radius	Overlap	Over max	Score
1	[45.66, 40.56, 810.52, 663.2]	659.19916560457	11.712630436924	yes	4
2	[3.5377912542611, 122.88340464607, 832.85177432791, 102.2142227637]	154.21666004677	6.3919845832792	yes	0
3	[2.5695947048491, 55.544483105892, 626.76977100429, 818.4035867151]	816.98189370543	6.3486551611765	yes	5
4	[27.187694470002, 7.383584927742, 834.12094957612, 165.16710031129]	171.35523483706	1.6998396082232	yes	0
5	[2.4970453904292, 107.56920206545, 147.79921302734, 600.96580139848]	754.27876623534	1.4457514612364	yes	9
6	[-333.39525695325, -33.332929255992, -551.13755538503, 1176.2567277667]	636.717573631	1.0569522747208	yes	8
7	[41.257027357444, 43.532006533725, 972.66947431812, 415.34266049214]	412.06540262312	12.106804100201	yes	2
8	[-246.31283775086, 22.874262963601, -500.71681310562, 933.80196469498]	665.99707917402	0.63332636294142	yes	5
9	[89.731672862793, 90.455627923196, 650.94499861178, 713.53201403252]	719.01647151605	0.8099069742284	yes	9
10	[94.182276086748, 244.90184263729, -312.34914808896, -15.984607496821]	496.19048469929	6.9579249445827	yes	4

Fig. 13. The first generation of detectors with unscaled values.

#	Centre	Radius	Overlap	Over max	Score
1	[-333.39525695325, -33.332929255992, -551.13755538503, 1176.2567277667]	754.27876623534	0.16176065722673	yes	3
2	[-246.31283775086, 22.874262963601, -500.71681310562, 933.80196469498]	496.19048469929	1.8652038367886	yes	2
3	[94.182276086748, 244.90184263729, -312.34914808896, -15.984607496821]	496.19048469929	6.9579249445827	yes	4
10	[-246.31283775086, 22.874262963601, -500.71681310562, 933.80196469498]	754.27876623534	0	no	0
9	[-333.39525695325, -33.332929255992, -551.13755538503, 1176.2567277667]	496.19048469929	0	no	0

Fig. 14. The fourth generation of detectors after running several suppressions.

#	Antigen	Result	Generation #
1	[60.06, 126.02, 978.89, 427.11]	Alarm!	1
2	[94.81, 125.18, 859.17, 743.68]	Alarm!	1
3	[52.6, 107.84, 421.43, 204.17]	Alarm!	1
4	[42.88, 9.89, 700.14, 28.52]	Alarm!	1
5	[10.83, 146.65, 80.78, 34.18]	Alarm!	1
6	[69.71, 141.42, 182.66, 172.82]	Alarm!	1
7	[17.43, 5.43, 377.09, 72.29]	Alarm!	2
8	[49.87, 143.89, 156.74, 168.58]	Alarm!	2
9	[8.02, 122, 819.52, 693.07]	Alarm!	2
10	[75.09, 60.43, 344.95, 944.64]	Alarm!	2
11	[1.85, 41.12, 91.18, 882.67]	Alarm!	2
12	[25.76, 6.93, -52.54, 313.08]	Alarm!	3
13	[82.26, 83.93, -34.98, 594.2]	Alarm!	3
14	[32.93, 40.11, 100.63, 897.77]	Alarm!	3
15	[82.7, 35.2, 952.59, 465.22]	OK	3
16	[86.71, 112.21, 250.6, 758.03]	OK	3
17	[5.75, 65.84, 106.16, 465.41]	OK	4
18	[18.47, 73.92, 604.71, 450.33]	OK	4
19	[78.26, 3.03, 679.72, 340.8]	OK	4
20	[90.77, 148.39, 265.3, 284.15]	OK	4

Fig. 15. Test runs with a sample of antigens on each detector generation with detection results.

The testing of the algorithms on a 2-dimentional space proves that the detectors show good coverage of the non-self space and a stable detection of non-self antigens. Fig. 13. shows the coordinates, radii, overlap and detection score of the first detector generation. The population should stay the same but after 3 generations the detector population decreased (Fig. 14.) but still detected the pathogens (Fig. 15). The chosen actions did not differ much probably because of the implementation which needs further research and improvement.

6.3 Computer simulation of the railway station

For the experiments the program for programmable controller was implemented. The controller performs all the calculations and controls the electric drive and traffic lights on the functional prototype.

The computer model is created to show the results of controller's operations to perform an emergency stop before the red signal of the traffic lights.

The specific environment is developed by the authors for the modelling of railway system for safety improving algorithms (Fig. 16).

The data from the specific memory addresses of the controller is read by the server (Fig. 17.) and transferred to the model.

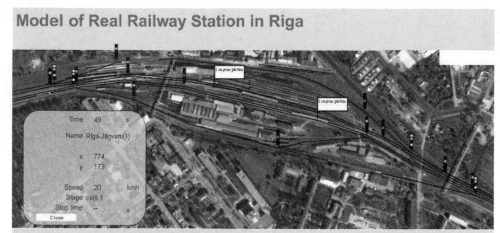

Fig. 16. Simulation environment

Name ⛟	Address	Data Type	Access
XL2	VD104	REAL	RW
XL1	VD100	REAL	RW
Xb	VD112	REAL	RW
Xa	VD108	REAL	RW
Speed	VD26	REAL	RW
L2YELLOW	Q0.6	BOOL	RW
L2RED	Q0.5	BOOL	RW
L2GREEN	Q0.7	BOOL	RW
L1YELLOW	Q0.3	BOOL	RW
L1Red	Q0.2	BOOL	RW
L1GREEN	Q0.4	BOOL	RW

Program23_4_2010
What's New
MicroWin(TCP/IP)
NewPLC

Fig. 17. Input data from PLC to computer model

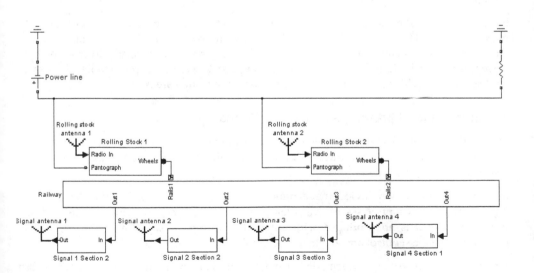

Fig. 18. Model of the railway system

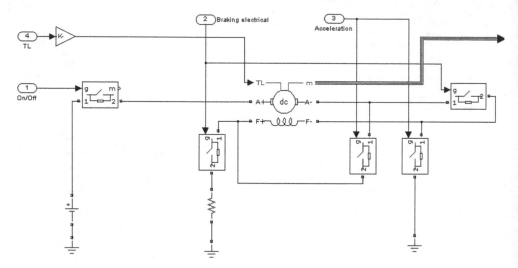

Fig. 19. Fragment of the electrical part of the computer model of the rolling stock

The current experiment is proposed for modelling of crash prevention of two trains moving towards each other. The model consists of 3 series block-sections; 2 rolling stocks; 4 railway signals (Fig. 18.).

Each rolling stock and signal is equipped with receiving and transmitting devices that give a possibility in a multi-agent system.

Electrical part of the model (Fig. 19.) consists of a DC drive with characteristics of 8 DC motors, 1 switch to connect or disconnect the electric drive from the electric contact network, and 2 pairs of switches for acceleration and for braking that changes direction of field current I_f flow. A braking branch contains braking resistance. An output of a DC drive is an electrical torque which handles the mechanical part of a rolling stock.

7. Experiments of prototype in real conditions

The result of this work is a train emergency braking device. The invented device is proposed to increase safety on railway transport. It gives possibility to stop rolling stock automatically before a closed signal timely.

In contrast to the known devices that actuate brake only after the passing of a closed signal, the invented device provides a train emergency braking and stopping before a closed section, even if it is not equipped with automatic locomotive signalling. The device also provides a distance control and an emergency braking way calculation.

The detector of the regular braking distance determines an emergency braking distance, but the detector of the starting point of regular braking defines the point on the route when braking should be started. The module for checking the reaching the starting point of regular braking evaluates location of the train, defines starting point of regular braking and operates braking signalling device in the cabin, the module for controlling starting of regular braking evaluates whether the regular braking is started. The device warns the driver about the necessity of starting process of the emergency braking taking into account the signal of the traffic light and speed limitation and allows to perform an automatic operation of the emergency braking in time and stops the train preventing trains collision at any sector of a railway.

Fig. 20. presents the demonstration of this device that can be installed on the train. Two traffic lights; the electric motor; sensors and wireless communication equipment are installed on the demonstrator.

According to the traffic light signal the controller selects the appropriate engine speed. When the red light is on, the control system automatically stops the engine. In response to the light sensor, the control unit in addition to the fan is turned on and switches to another mode of operation. Remote monitoring and control of the processes is possible using wireless communication. In a real system, it could be dispatching control centres, from which it is possible to switch both signals and also take over control of the train speed.

Taking into account the pieces of advice and the recommendations from the State joint stock company "Latvian Railways" (Latvijas Dzelzceļš/LDz) specialists, the prototypes of the locomotive and the signal devices have been created. Both inventions were issued Latvian and International Patents No. LV13978 B, LV14156 B, LV14187 B, WO 2011/115466 A2, PCT/EP2011/067474.

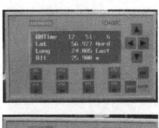

Fig. 20. Functional prototype and information screen with satellite navigation data

The authors and the LDz staff had tested the prototypes of the devices in real service conditions. A non-busy section of the railway was chosen to play the role of a proving ground. During the experiment all the devices were working steadily and without troubles, thus the experiment proved that the ideas adopted in the devices can be implemented into practice.

The task of the locomotive's embedded SAFE-R 3 device and the traffic lights' embedded SAFE-R 4 device, which was designed by the RTU and LDz, is to stop the train automatically at the restrictive signal of the traffic lights, in those cases when a driver does not react to this restrictive signal. It is provided that these devices will also work in unencoded railway sections, where the automatic locomotive signalling did not work.

8. Conclusions

Advantages of the proposed device are the following: The device is not using rail circuits and works independently of automatic locomotive signalization system. The proposed device is an alternative or auxiliary to existing safety systems. As opposed to existing systems the new device uses wireless communication network and may work in railway sections without an automatic interlocking system. The possibility to prevent a dangerous situation and a crash corresponding to the condition of the braking system of rolling stock allows stopping the train before dangerous failure time point; possibility of using of already existing measurement devices and sensors together with the new sensors.

The results of the experiment show the possibility to use the proposed system as an auxiliary safety device to prevent breaches of red signal and crashes on the railway.

The most relevant features of immune algorithms are self-learning, diversity maintenance, memory about the past decisions, and detection of previously unknown but related elements, noise rejection and classifying ability.

An intelligent rolling stock safety control system could benefit from using a combination of both an immune negative selection algorithm and a clonal selection algorithm. A fault detection system for railway electric transport could benefit from using an immune negative selection algorithm.

The most feasible way to implement a railway electric transport safety control system would be through the two phases of anomaly detection and determination of their type to draw a conclusion about further action.

Single string data encoding is better suited for use on PLC. The PLC program needs a data buffer to eliminate the risk of data loss due to unstable radio signal.

The authors need to assess the possibility to run the data analysis using these algorithms in real time. The detector maturation and control cell selection processes need improvement.

9. References

Russel S. J., Norvig P.. Artificial Intelligence. A Modern Approach, 2 edition. Prentice Hall, 2006, 1408p.

Luo R., Zeng J. Computer simulation of railway train braking and anti-sliding control. //In Proceedings of 21st International Symposium on Dynamics of Vehicles on Roads and Tracks (IAVSD'09), 2009 – 189 p.

Hasegawa Y., Tsunashima H., Marumo Y., Kojima T. Detection of unusual braking behavior of train driver //In Proceedings of 21st International Symposium on Dynamics of Vehicles on Roads and Tracks (IAVSD'09), 2009 – 166 p.

Gorobetz M. Research of Genetic Algorithms for Optimal Control of Electric Transport. Promotional thesis. Riga, 2008 - 189 p.

Levchenkov A., Gorobetz M., Ribickis L., Balckars P.   „Generating of Multi-Criteria Alternatives for Decision-Making in Electric Light Rail Control" //In China-USA Business review, December 2009, pp. 49-55.

Forrest S., Perelson A.S., Allen L., Cherukuri R. Self-nonself discrimination in a computer. In proceedings of the 1994 IEEE Symposium on Research in Security and Privacy. Los Alamitos, CA.

Kephart J.O.. A biologically inspired immune system for computers. In proceedings of Artificial Life IV: The Fourth International Workshop on the Synthesis and Simulation of Living Systems, 1994. MIT Press.

Dasgupta D., Krishna Kumar K., Wong D., Berry M.. Negative Selection Algorithm for Aircraft Fault Detection. In proceedings of ICARIS 2004.

Mor-Yaroslavtsev A., Levchenkov A. Rolling Stock Location Data Analysis Using an Immune Algorithm on an Intelligent Embedded Device. // In proceedings of 19th Telecommunications Forum TELFOR 2011.

Powers S. T., He J.. A hybrid artificial immune system and Self Organising Map for network intrusion detection. Elsevier IS 179, 2009.

Musilek P., Lau A., Reformat M., Wyard-Scott L.. Immune Programming. Elsevier IS 179, 2009.

Gorobetz M., Alps I., Levchenkov A.. Mathematical Formulation of Public Electric Transport Scheduling Task for Artificial Immune Systems. Proceedings of ITELMS '2009, Kaunas.

Masutti, T. A. S. Castro L. N. de. A self-organizing neural network using ideas from the immune system to solve the traveling salesman problem. Elsevier IS 179, 2009.

Tavakkoli-Moghaddam R., Rahimi-Vahed A., Mirzaei A. H.. A hybrid multi-objective immune algorithm for a flow shop scheduling problem with bi-objectives: Weighted mean completion time and weighted mean tardiness. Elsevier IS 179, 2009.

PCT/EP2011/067474. Device for Safe Passing of Motor Vehicle over Level Crossings Using Satellite Navigation Systems. A.Ļevčenkovs, M.Gorobecs, I.Raņķis, L.Ribickis, P.Balckars, A.Potapovs, I.Alps, I.Korago, V.Vinokurovs, 6.10.2011. (26.07.2011.)

WO 2011/115466 A2, (PCT/LV2011/000004) Controlling Device of Railway Track Sections. A.Ļevčenkovs, M.Gorobecs, J.Greivulis, P.Balckars, L.Ribickis, I.Korago, A.Bobeško. 22.09.2011. (17.03.2010.)

Patent application Nr. P-11-102. Device for Safe Passing of Motor Vehicle over Level Crossings Using Satellite Navigation Systems. A.Ļevčenkovs, M.Gorobecs, I.Raņķis, L.Ribickis, P.Balckars, A.Potapovs, I.Alps, I.Korago, V.Vinokurovs, 26.07.2011.

Patent application Nr. P-11-76. Train anticollision device with satellite navigation. A.Ļevčenkovs, M.Gorobecs, I.Raņķis, L.Ribickis, P.Balckars, A.Potapovs, 23.05.2011.

Patent Nr. LV13978 B. Train Emergency Braking Device. Gorobecs M., Greivulis J., Ļevčenkovs A., Balckars P., Ribickis L. 14.05.2009.

Patent Nr. LV 14156 B. Controlling Device of Railway Track Sections. A.Ļevčenkovs, M.Gorobecs, J.Greivulis, P.Balckars, L.Ribickis, I.Korago, A.Bobeško. 17.03.2010.

Patent Nr. LV 14187 B. Train's Braking Way Control Device. A.Ļevčenkovs, M.Gorobecs, J.Greivulis, I.Uteševs, P.Balckars, L.Ribickis, V.Stupins, S.Holodovs, I.Korago

## 3

General Principles Regarding the Rehabilitation of Existing Railway Bridges

Petzek Edward and Radu Băncilă
"Politehnica" University of Timișoara & SSF-RO Ltd
Romania

1. Introduction

Rehabilitation and maintenance of existing steel constructions, especially steel bridges is one of the most important actual problems [1],[2],[3],[8],[14]. The infrastructure in Romania and in other East – European countries has an average age of about seventy to ninety years. Many of these structures are still in operation after damages, several phases of repair and strengthening. Replacement with new structures raises financial, technical and political problems. The budget of the administration gets smaller. The present tendency to raise the speed on the main lines to a level of v ≤ 160 km / h must be emphasized (Figure 1).

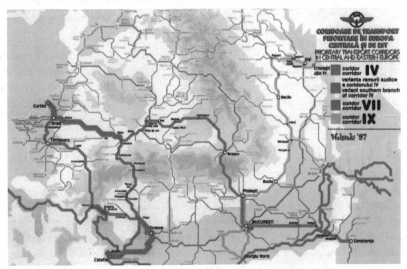

Fig. 1. The European corridors crossing Romania's territory: IV, VII and IX

During service, bridges are subject to wear. In the last decades the initial volume of traffic has increased. Therefore many bridges require a detailed investigation and control. The examination should consider the age of the bridge and all repairs, the extent and location of any defects etc (Figure 2).

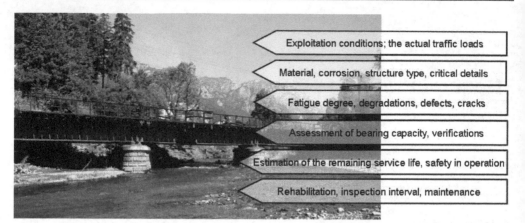

Fig. 2. Assessment and control of existing steel bridges

Carefully inspection of the structure is the most important aspect in evaluating the safety of the bridge. On the accuracy of the in situ inspection depends the level of evaluation.

The check of existing structures should be based on the complete bridge documentation (drawings with accuracy details, dimensions and cross sections of all structural elements, information about structural steel, stress history. However, in many cases these documentations are incomplete or missing. But these informations can be recovered due to the carefully investigations and inspections of the structures, experimental determination of the material characteristics and stresses in structural elements, full scale in situ tests (static and dynamic), calibration of structure and spatial static analysis.

Today, the budget of the administration and the owners (i.e. the railways and highway companies) get smaller. In consequence it is necessary to invest the available money where there will be the greatest benefit. Therefore, those responsible for the decisions need information about the safety of the structure, the remaining life, the costs for maintenance etc. Nobody will take the responsibility for failure of a structure as a result of budget restrictions.

Bridge life is generally given by fatigue; difficult is the estimation of the loading history. For bridges where the stress history is known the fatigue life may be calculated using the Miners's rule and an appropriate S-N curve; also the assumption of the same spectrum for bridge life (or for certain periods) must be made. During the process of assessment the fatigue life of old riveted bridges, is important to establish the proportion of the whole fatigue life that has been already got through. For stringers and cross girders of existing railway bridges the number of 10^7 cycles is exceeded. It might be affirmed that for such bridges if no fatigue cracks can be detected, no fatigue damage has occurred! Subsequently, if the loading spectrum remains the same in the future as in the past, fatigue cracking might not take place! In almost all the cases the loadings increased; in this situation survival for 100 years (or more) without cracking would not justify the assumption that no damage has yet occurred! Minor cracking is difficult to detect during usual inspections.

2. Technical condition of existing bridges

Rehabilitation and maintenance of existing steel bridges is one of the most important actual problems.

A continuous maintenance, which generally must increase in time, is important in order to assure the safety in operation of the existing structures.

The evaluation of the current technical condition of an existing steel bridge structure with the help of in situ evaluation depends, in a high percentage, on the engineer's qualification. In the case of in situ inspection it is recommended to insist on the appeared fatigue defects, of riveted or welded connections (the expert must to insist on the connections between the stringers with cross girders and cross girders with main girders), critical details (which are included in standards or catalogues), corrosion level, the structure deformations due to traffic, bridge bearings. The expert can use non-destructive techniques to determine the integrity of base material or structural components. The non destructive testing of the inspection of in service damage in bridge elements are the following:

- *visual inspection* – the most common method which includes microscopes, mirrors, portable video cameras, robotic crawlers; this method is very useful in case of surface cracks;
- *magnetic particle inspection* – this method is also very simple and does not need high qualification personnel, but can be applied just in case of ferromagnetic materials (not for austenitic steels). The method consists in the magnetization of the high stress elements or critical details and indicates directly the surface discontinuity through forming a distorted magnetic field, which can be detected under proper lighting conditions;
- *liquid penetration inspection* – is a simple method including the qualification of the personnel; it uses penetrate liquids with fluorescent pigment and UV – light in order to indicate the surface defections;
- *radiographic inspection* – the method is applied for hidden defects and it uses Gamma or Roentgen radiation. The inspected element is placed between the radiation and the film. The interpretation of the radiographic images should be done by experts, otherwise defects could be ignored;
- *ultrasonic inspection* – this testing is used for flaws and cracks in the material thickness, on the surface or hidden defects; highly qualified personnel is needed. High frequency sound waves are introduced into a material and they are reflected back from surfaces or flaws. This process is recorded by an oscilloscope. This method cannot be used for elements made of multiple plates (riveted sections).
- *Eddy Current testing* – this method can detect surface defects but can also be used for thickness inspection.

The usual simplified analysis methods do not always give the lowest resistance values for the structure, but usually the more refined assessment methods which give greater resistance are expensive. According to the experience of the expert a progressive analysis can be applied. In a first step simple classical methods can be applied [2]. If they fail, more sophisticated methods can be used, until either it is shown that the bridge is adequate, or it is concluded that strengthening is needed. An engineer with experience can jump over some time consuming steps which do not give any benefit (an interesting proposal is that the engineer will be paid on a percentage on the saving basis [4]). Actual loads are lower than those used for design purposes. Fatigue tests on elements taken from demolished structures gives – generally - greater fatigue life than the values according the codes [5].

The applied stress range, the geometry of the detail and the number of stress cycles has a decisive effect on the remaining fatigue life of the structures.

By differences of more than 5 % of the cross section – due to corrosion, the actual values must be introduced.

However, from the overall examination of a large number of bridges many defects can be pointed out. The defects are widespread, having a heterogeneous character from the point of view of location, development and development tendency; their amplification was also due to the climate and polluting factors that caused the reduction of the cross section due to corrosion. Statistically, in 283 from among 1088 welded bridges, and in 356 from among 3201 steel riveted bridges cracks were detected and repaired. It is not allowed to weld cracks! Old bridges can have welds executed in the early years; a special attention must be paid to these parts. Generally the riveted connections have a good behavior in time due to the initial pre-stressing force which can reach 70 – 80 N/mm². In Table 1 some typical defects in stringers, cross girders, main girders wind bracings and orthotropic deck and their repair are presented [6],[7].

In figure 3 there is presented a crack of a joint plate from a wind bracing and in figure 4 also cracks in the lower flange joint of a double T girder, near to the bearing.

Fig. 3. Crack in the joint plate

Fig. 4. Crack in the lower flange near to the bearing

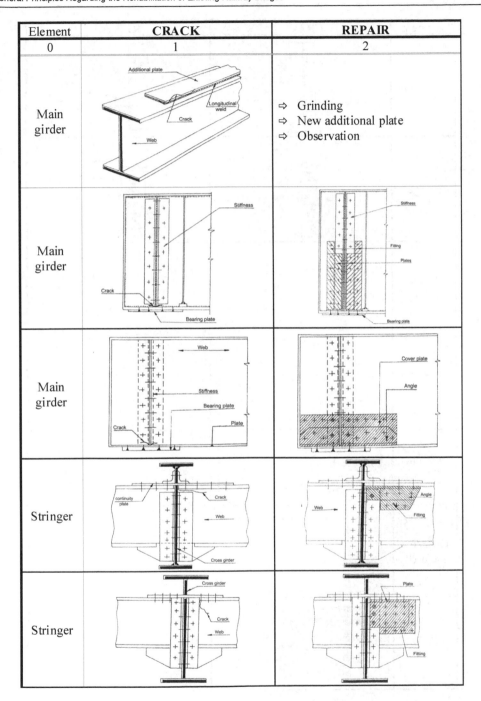

Element	CRACK	REPAIR
0	1	2
Main girder		⇨ Grinding ⇨ New additional plate ⇨ Observation
Main girder		
Main girder		
Stringer		
Stringer		

Table 1. Typical defects in welded and riveted steel bridges

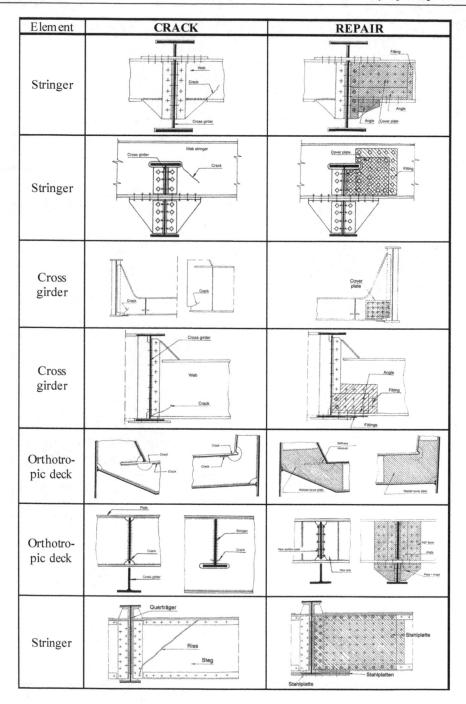

Element	CRACK	REPAIR
Stringer		
Stringer		
Cross girder		
Cross girder		
Orthotro-pic deck		
Orthotro-pic deck		
Stringer		

Table 1. Typical defects in welded and riveted steel bridges - continuation

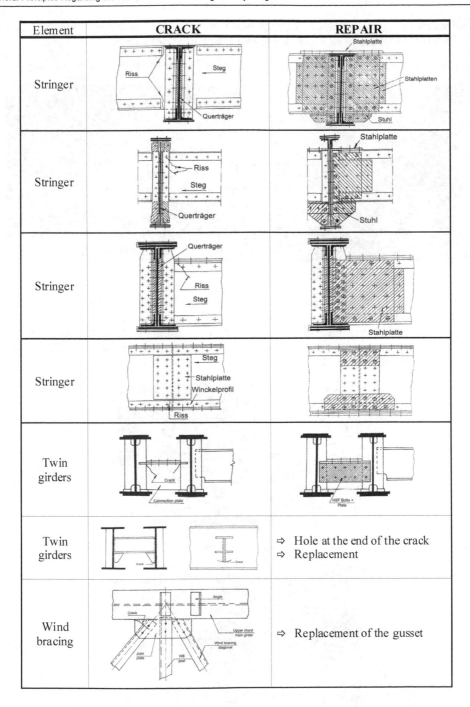

Table 1. Typical defects in welded and riveted steel bridges - continuation

Signs of cracks and defects are rust traces, which occur by friction between jointed plates of the elements. These are of relevance for hidden constructive elements.

Defects of the bearings are frequent as well. In figure 5 there are presented two examples.

Fig. 5. Defects on bearings

Due to deficient maintenance the riveted structures are strongly corroded, especially in the lower zones (figure 6). These corroded surfaces represent also a critical detail for fatigue. Fatigue under corrosion factor is an aspect treated only qualitatively and not quantitatively. The interaction between the two aspects is obvious. There can also found inefficient joints, like weakened rivets (figure 7).

Fig. 6. Strong corrosion at the lower flange – railway bridge on a main line

Fig. 7. Inefficient joints – railway bridge on a main line

3. Characteristics of materials

The following facts show that a material analysis for old riveted bridges is very useful:

- Old bridges are in many cases erected using material with very poor welding qualities and basing on railway administration data and specialized literature it is known that cast iron was used to build bridges;
- The specialized literature doesn't offer enough information about this structural steel;
- The structural material comes from several producers (for South Eastern Europe mostly from Reschitz – Romania and Győr – Hungary).

Span	Year of construc-tion	Position	Type of structures
101,76 m	1912 (St.E.G. Reschitz)	Mehadia	
81,6 m	1911 (St.E.G. Reschitz)	Valea Cernei	
71,76 m	1913 (St.E.G. Reschitz)	Balta Sărată	
50,65 + 66,20 + 36,70 + 2x30,40 m	1912 (St.E.G. Reschitz) replaced 1997	Şag	
2x76,80 + 3x51,42 m	1912 (Győr) replaced 2000	Arad	
56,2 + 2x31,0 + 4x 32,7 m	56,2 m 1927 (Reschitz) 2x31,0 + 4x32,7 m 1907	Branişca	

* Possible construction year

Table 2. The bridges on which the present material study basis

The presented study's results can be extended to Middle and South Eastern Europe when the history of communication ways and the state of old railway and highway steel bridges in this region is regarded. In this context the following event can be mentioned: on the 1st of January 1855 the "Kaiserliche und Königliche Privilegierte Österreichische Staatseisenbahngesellschaft" (St.E.G.) took over all steel producers in Banat. The investments in Reschitz (in present Resita) turn the steel mill into an important bridges' factory. The production of steel bridges reached 3960 tonnes in 1910. Bridges made by St.E.G. Reschitz are still in use in Romania, Austria and Hungary. Between 1911 – 1913, 1620 tonnes of bridge structures made of cast iron were replaced in the western part of Romania (Banat), namely on the railway segment Timişoara – Orşova. In this sense the material study took into account bridges from this region, built around 1911. Following material analysis were performed in order to determine the characteristics of the material: chemical analysis, metallographic analysis, tensile tests, Brinell tests, Charpy "V Notch" tests. The samples were taken from secondary elements, but also in some cases (Bridges in Arad and Şag which were replaced) from main elements: stringers, cross girders, main girders [7]). The results of the chemical analysis are presented in the following table.

No.	Bridge / Specimen number & position	Chemical composition						
		C %	S %	Mn %	P %	Si %	Ni %	N %
1.	Valea Cernei Bridge / S1-VCB / Secondary elements	0,12	0,034	0,39	0,021	0,023	-	-
2.	Valea Cernei Bridge / S2-VCB / Sec. Secondary elements	0,11	0,020	0,32	0,024	0,010	-	-
3.	Mehadia Bridge / S1-MB / Secondary elements	0,11	0,014	0,50	0,028	0,050	-	-
4.	Balta Sărată Bridge / S1-BSB / Secondary elements	0,07	0,072	0,38	0,016	0,010	-	-
5.	Şag Timiş Bridge / S1-STB / Secondary elements – Span I	0,16	0,058	0,46	0,035	0,035	-	-
6.	Şag Timiş Bridge / S2-STB / Secondary elements – Span II	0,14	0,066	0,63	0,030	0,061	-	-
7.	Şag Timiş Bridge / S3-STB / Secondary elements – Span III	0,13	0,060	0,50	0,051	0,112	-	-
8.	Şag Timiş Bridge / S4-STB / Secondary elements – Span IV	0,18	0,054	0,48	0,066	0,056	-	-
9.	Şag Timiş Bridge / S5-STB / Secondary elements – Span V	0,18	0,017	0,44	0,059	0,010	-	-
10.	Brănişca Bridge / S1-BB / Secondary elements	0,12	0,070	0,39	0,013	0,060	-	-
11.	Brănişca Bridge / S2-BB / Secondary elements	0,11	0,055	0,38	0,014	0,060	-	-
12.	Arad Bridge / S1-AB / Secondary elements – Span I	0,14	0,061	0,53	0,038	0,069	-	-
13.	Arad Bridge / S2-AB / Secondary elements – Span II	0,19	0,033	0,64	0,051	0,018	-	-
14.	Arad Bridge / S3-AB / Main elements – Stringers Span III	0,089	0,032	0,531	0,009	0,018	0,067	-
15.	Arad Bridge / S4-AB / Main elements – Cross girders Span III	0,058	0,059	0,485	0,017	0,018	0,037	-
16.	Arad Bridge / S5-AB / Main elements – Main girder Span III	0,056	0,032	0,493	0,001	-	0,031	-
17.	Arad Bridge / S6-AB / Secondary elements – Span IV	0,18	0,035	0,46	0,063	0,078	-	-
18.	Arad Bridge / S7-AB / Secondary elements – Span V	0,1	0,047	0,43	0,020	0,030	-	-
19.	St 37 (STAS 500/2-80)	0,25	0,065	0,85	0,065	0,07	0,30	0,015
20.	St 34 (STAS 500/2-80)	0,17	0,055	0,60	0,055	-	-	-
21.	INCERTRANS Bucharest Research for cast iron	0,04... 0,11	0,014...0, 043	0,15... 0,48	0,121...0, 32	0,07... 0,31	-	-
22.	German Research for cast iron*	0,16	0,056	0,100	0,470	0,100	0,007	-

* German research – Stahlbau 05.1985 (Prof.Dr.Ing. D. Kosteas, Ing. W. Stier, Ing. W. Grap)

Table 3. Chemical analysis results

The statistical interpretation of the tensile tests results shows a minimal value for the yield stress of 230 N/mm².

The impact tests on Charpy V Notch specimens lead to conclusion that the transition temperature is situated in many cases in the range from -10°C to 0°C (figure 8).

By analyzing the laboratory results it can be concluded that the steel is a mild one, that could be associated to the present steel types St 34 or St 37.1 (S 235 according to the Eurocodes). Also, on the two dismantled bridges – Arad and Şag – fracture mechanics tests were made [7] in order to establish the integral value J_c according to ASTM E813-89 (figure 9), the CTOD and to determine the fatigue crack growth rate and the material constants C and m according to ASTM E647-93. For these tests compact specimens CT (thickness 8 mm) as well as bending specimens for CTOD, have been used.

They have been obtained from the stringers, cross girders and main girder – lower chord. The minimal value of material toughness in term of J-Integral for these old riveted steel bridges is J_{crit} = 10 ... 20 N/mm for a temperature of -20°C.

The method of fatigue assessment for structural elements with defects, was developed basing on the possibility of crack propagation modeling under fatigue loads and with the help of known laws.

The method is founded on the recommendations of the BS 7910:1999.

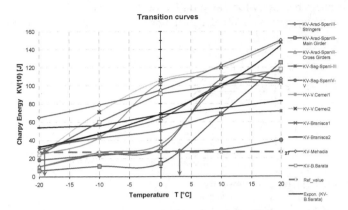

Fig. 8. Transition curves for the analyzed bridge structures

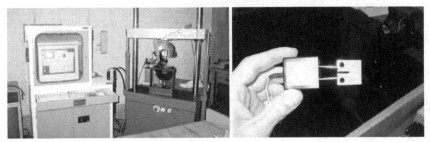

Fig. 9. Experimental tests for determining of J values

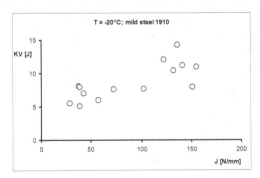

Fig. 10. Charpy V Notch Energy vs. material toughness J_c (Arad Bridge 1912)

In the present state of knowledge it is generally accepted that the fatigue failure of materials is a process containing three distinct steps: (1) initiation of defect (crack), (2) crack propagation in material, (3) separation through complete failure of the material in two or more pieces. Practically, the safety service life of an element under fatigue conditions can be expressed as follows (figure 11):

$$N_f = N_i + N_p \tag{1}$$

N_i = number of cycles necessary for the initiation of the defect (crack)
N_p = number of cycles necessary for the propagation of the defect until the occurrence of failure.

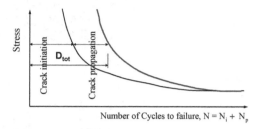

Fig. 11. Fatigue life of structural elements

The evaluation of crack propagation conditions can be accomplished with the help of characteristically values, which are founded on fracture mechanics concepts: material toughness express by the stress intensity factor K or J integral value and the crack growth rate da/dN (crack growth for each load cycle). A relation of the following type can express the crack growth rate (Figure 12):

$$\frac{da}{dN} = f\left(\Delta K, R, H\right) \tag{2}$$

da/dN - crack extension for one load cycle; ΔK- stress intensity range, established basing on the stress range $\Delta\sigma$; R- stress ratio; H- indicates the stress history dependence.

The crack growth rate da/dN, defined as a crack extension - da obtained through a load cycle dN (it can also be defined as da/dt, in which case the crack extension is related to a time interval), represents a value characteristic of the initiation phases respectively the stable crack propagation. It has been experimentally observed that the connection between the crack growth rate and stress intensity factor variation represents a suitable solution for the description of the behavior of a metallic material containing a crack, as in the case of steel. In a logarithmic graphical representation of the crack growth rate da/dN versus the stress intensity range ΔK a curve as the one in the following figure (Fig. 12), is obtained.

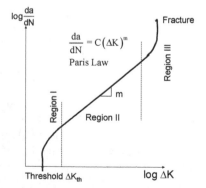

Fig. 12. Logarithmic representation of the fatigue crack growth in steel

The calculation of the structural elements remaining service life can be done basing on the Paris law, more precisely by integration of this law:

$$N = \int_0^N dN = \int_{a_0}^{a_{crit}} \frac{da}{C \cdot \Delta K^m}$$ (3)

N - number of stress cycles necessary in order that the crack extends from its initial dimension a_0 to the critical value a_{crit}, where failure occurs;
a – crack length;
C, m – material constants from the crack propagation law;
ΔK – stress intensity factor range.

This integral can be numerically calculated by taking into account a critical detail knowing the crack values (initial and critical), basing on the following relation:

$$N = \int_{a_0}^{a_{crit}} \frac{da}{C \cdot \Delta\sigma^m \cdot Y^m \cdot (\pi a)^{-m/2}}$$ (4)

The number of cycles Ni obtained with the help of relation (4) represents the remaining service life of the detail, by regarding the initial length a_0 up to the critical length a_{crit}, by admitting stable crack propagation (Figure 13).

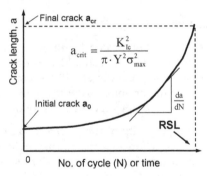

Fig. 13. Principles for determining the remaining service life

The critical crack value can be calculated basing on the K criterion respectively on the J or CTOD criterions or with the help of the failure assessment diagram.

The C and m material constants from the Paris law are experimentally determined by fracture mechanical tests. In this sense in most cases compact specimens C(T), three point bended specimens SEN(B) and middle central panels M(T) are used. Such a standard which describes the test methods for the determination of the crack growth rate, is the American Standard ASTM E 647 (Standard Test Method for Measurement of Fatigue Crack Rates).

Basing on the determined values da/dN and ΔK the program presented below, also automatically determines the C and m material constants by the Paris relation:

$$\ln(\frac{da}{dN}) = \ln C + m \cdot \ln \Delta K$$ (5)

The experimental tests on 26 CT specimens from two old bridges have shown that for the oldest mild steels the values of the material constants from the Paris relation are in the following intervals: m = 2,05 … 5,65 and C = 2,2 x 10^{-11} … 10^{-18}.

Relatively large value of m corresponding to very small values of C, for example for m > 4 → C ≅ 10^{-15} … 10^{-18}.

Fig. 14. Experimental tests in order to establish the material constants C and m

4. Present verification concept

During service bridges are subjected to repeated loadings causing fatigue. Therefore many bridges require an inspection. The examination should consider the age of the bridge and all repairs, the extent and location of any defects etc. [8]. A continuous maintenance, which generally must increase in time, is important in order to assure the safety in operation of the existing structures. The present methodology includes the following stages [9]:

Step 1. estimation of the loading capacity of the structure based on a detailed inspection; analysis of drawings, inspection reports, repairs, reinforcements, analysis of the general behavior of the bridge (displacements, vibrations, corrosion and cracks). In this phase the stresses in the structure can be calculated with the usual simplified hypothesis;

Step 2. the accurate determination of the stresses in the structure and of the remaining safety of the elements. This phase includes: tests on materials, computer aided analysis of the space structure, remaining safety calculated on the base of the real time - stress history;

Step 3. in situ static and dynamic tests.

This methodology adopted by the Romanian standard is illustrated in Figure 15.

The calculation of remaining fatigue life is normally carried out by a damage accumulation calculation. The cumulative damage caused by stress cycles will be calculated; failure criteria will be reached.

$$D = \sum \frac{n_i}{N_i} \leq 1 \tag{6}$$

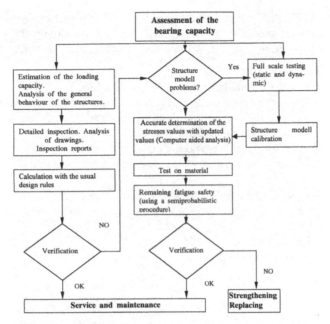

Fig. 15. Methodology of the Romanian standard SR 1911-98 [3]

The classical fatigue concept is based on the assumption that a constructive element has no defects or cracks. However, discontinuities and cracks in the components of structures are unavoidable, basically because of the material fabrication and the erection of structures. It is very clear that the kind of fatigue cracks, which are initiated by structural non-homogeneity (possible non-metallic inclusions or other impurities), surface defects (including corrosion) and the stress factor, are present in the old riveted structures.

The presence of cracks in structural elements modifies essentially their fracture behavior. Fracture, assimilated in this case as crack dimensions growth process under external loadings, will be strongly influenced by the deformation capacity of material. The FM approach has acceleration in damage increase; with increasing damage a smaller stress range contribute to the damage increase. The authors proposed [9], a complementary method based on the fracture mechanics basic concept

$$J_I \leq J_{Ic} \tag{7}$$

in order to calculate the remaining fatigue life.

In practice two situations can be distinguished:

- D < 0,8 the probability to detect cracks is very low. The inspection intervals (generally between 3 – 6 years) can be established on criteria independent of fatigue. Nevertheless, a special attention must be paid to critical details.
- D ≥ 0,8 cracks are probable and possible. An in situ inspection and the analysis of critical details are strongly necessary. Also a fracture mechanics approach is recommended.

Generally, the establishing of the maintenance program, the determination of inspection intervals, the inspection priorities of structural elements and finally the calculation with high accuracy of the remaining service life of old riveted bridges takes into account the following main data:

- type of structure and exploitation conditions (traffic events);
- information about structural steel (mechanical properties – yield strength, tensile strength, hardness, transition curve ductile – brittle and transition temperatures, chemical composition, metallographic analysis);
- determination of critical members and details;
- crack detection and inspection techniques for evaluation of the initial crack size – a_0 and crack configuration;
- recording of the stress spectrum for the critical members under the actual traffic loads;
- evaluation of the critical crack size – a_{crit} based on failure assessment diagrams;
- fracture mechanics parameter – K_{crit}, δ_{crit}, J_{crit} (fracture toughness);
- simulation of the fatigue crack growth;
- temperature, environment conditions.

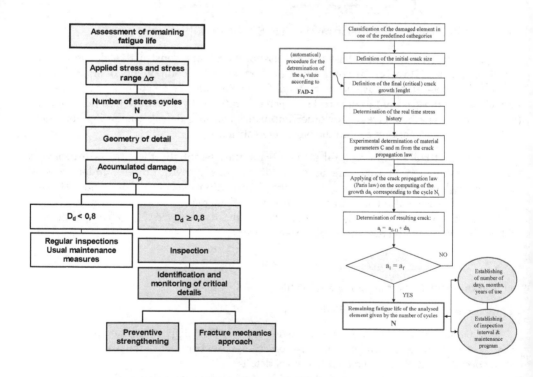

Fig. 16. Assessment of the remaining fatigue life and the crack growth procedure

Fig. 17. Soft for life prediction analysis

The methodology is conceived as an advanced, complete analysis of structural elements containing fatigue defects, being founded on fracture mechanics principles and containing two steps: namely one of determination of defects' acceptability with the help of Failure Assessment Diagrams (level 2) [10] and of determination of final acceptable values of defect dimensions; this is followed by a second step which in fact represents a fatigue evaluation of the analyzed structural elements basing on the present stress history recorded on the structure, on the initial and final defect dimensions and the FM parameters, namely the material characteristics C and m from the Paris relation (crack growth under real traffic stress) and further on the exact determination of the number of cycles N needed in order that a fracture take place, respectively the determination of the remaining service life of the structural elements (years, months, days).

In order to determine the remaining service life it is important to know how long it will take the crack to grow from the minimum detectable size to the critical value. In this situation the safe inspection intervals can be calculated with the following relation:

$$\Delta T_{insp} = N_{RFL} / N_{\Delta a},\qquad(8)$$

where

N_{RFL} = the remaining fatigue life calculated for the structural element
$N_{\Delta a}$ = the number of cycles computed for a crack extension rate of 5 mm during two successive inspections.

The life prediction computing was performed with the help of a soft developed by one of the authors [7].

5. Case studies

Three European corridors cross Romania's territory: IV, VII and IX. Of a special interest for many European countries is the Pan-European Corridor IV „Berlin - Nürnberg - Prague - Budapest - Constanta /Istanbul/ Thessaloniki". On Romania's territory the railway component of Pan - European Corridor IV has the following route: Curtici - Arad - Sighisoara - Brasov – Predeal – Campina - Bucharest - Constanta.

Due to the fact that the Campina - Predeal railway route crosses the sub-Carpathians area and the southern part of the Southern Carpathians it was necessary to adapt it to the difficult geographical conditions; actually this section is the most complicated part from the whole Romanian route. The railway line was built step by step beginning with 1879. The doubling of the 43 km long railway line Campina - Predeal was accomplished between 1939 - 1942. The line electrification was completed during 1961 - 1965.

Generally, the line is oriented from the south to the north. It follows the valley of the Prahova River crossing this river in 17 points.

This section includes 41 bridges. From this total number 22 are plate girder bridges, 6 are truss structures, 12 concrete bridges and 1 are conceived in the filler beams deck solution.

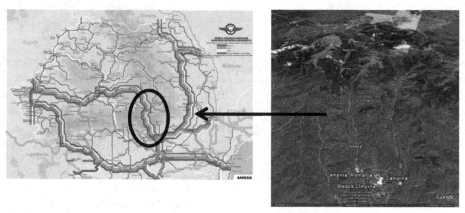

Fig. 18. Emplacement plan

All these structures were verified. In the following some aspects regarding the verification of the steel bridges are presented. The methodology which was adopted is showed in the figure 19.

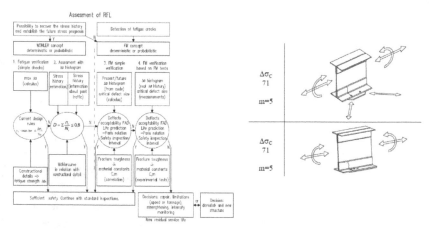

Fig. 19. The methodology and the constructive details for the evaluation of the RFL

After the stress analysis (Figure 20) and the fatigue verifications based on the Wöhler concept which were made in relation with the prescriptions of the Romanian standard SR 1911-1998, Swiss code 161 & SBB Richtlinie 2002 and the German code DS 805-2002, a life prediction analysis based on fracture mechanics principles was performed in order to evaluate the remaining fatigue life for these structures for the new traffic UIC conditions.

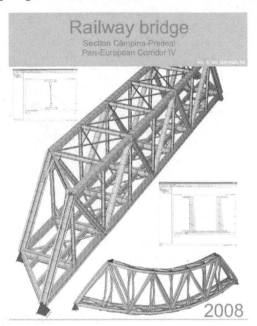

Fig. 20. The static model of the structure no. 6, km 108+690,34

For example, a plate girder bridge (no. 28, at km 125+323.25) is presented, which is a riveted structures crossing the Prahova River. The structure has four spans, one of 13.90 m and three of 14.10 m each and is skew (skew to the right - 48°37′) and its superstructure is made out of independent plate girder decks.

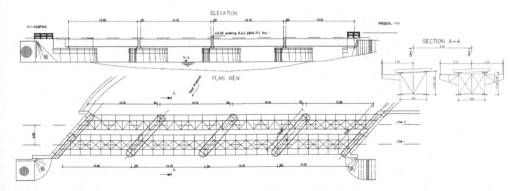

Fig. 21. The general disposition of the bridge no. 28

Calculation elements:

- Construction year – 1940.
- Due to lack of any documentation, the dimensions of the main girders elements were measured in the emplacement (Figure 22).

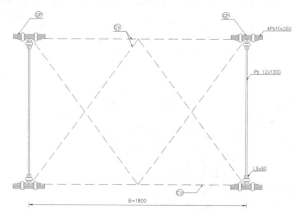

Fig. 22. Dimensions of the steel elements – Main girders

- The basic material to be considered is mild steel similar with St 37 (S235).
- The static scheme - simply supported girders.

The traffic on one main railway line was considered:

- the reference traffic 20-24x10⁶ t/ line/ year
- the Calculation Centre of SNCF CFR S.A. for the year 2004: 13,7 t/ line / year =>12-16 t / line/ year.

The traffic expressed in pairs of trains per day in the peak month that took place on the line Câmpina – Predeal in the year 2004, was:

- 28 pairs of passenger trains / day;
- 17 pairs of freight trains / day;

Total – 48 assimilated pairs of trains / day

For the main girders of the steel decks, all the fatigue checks were performed in the section from the middle of the span.

According to the stress analysis, the maximal stress range for UIC 71 convoy is:

$$\max \Delta\sigma_{UIC} = 97.13 \, N / mm^2$$

The resulted damage accumulated value (Miner rule) is D = 0.98.

Also, the complementary method of fracture mechanics was applied. For the material characteristics followings values were considered: the material is mild steel similar to the former steel St 34 - 37. n (Romanian standard - STAS 500/2 – 80); yield stress is σ_y = 230 N/mm²; tensile stress σ_{ult} = 360 N/mm².

For the material toughness in terms of J_{crit} a minimal value of 20 N/mm at a temperature of -20°C was chosen. For the life prediction procedure in the case of the material constants following values have been chosen: m = 3 and C = 3 x 10^{-12} (see also [11]).

A stress history was established using the following mixed traffic from EC 1. This is actually the future traffic which will be characteristic for the new rehabilitated railway corridor. All these trains were moved on the structure in order to establish the multi-block stress history.

Train type	No. of trains / day	Train weight [t]	Traffic volume [Mil. t/year]
1	12	663	2,90
2	12	530	2,30
3	5	940	1,72
4	5	510	0,93
5	7	2160	5,52
6	12	1431	6,27
7	8	1035	3,02
8	6	1035	2,27
	67		**24,95**

Table 4. Mixed traffic from EC 1

All these trains were moved on the structure in order to establish the multi-block stress history.

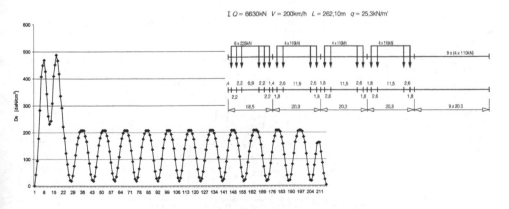

Fig. 23. The stress range diagram Δσ for train type 1

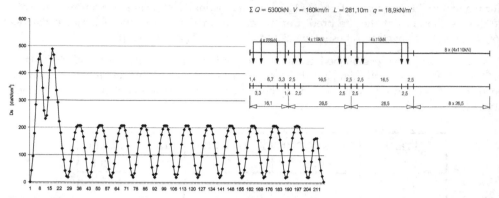

Fig. 24. The stress range diagram Δσ for train type 2

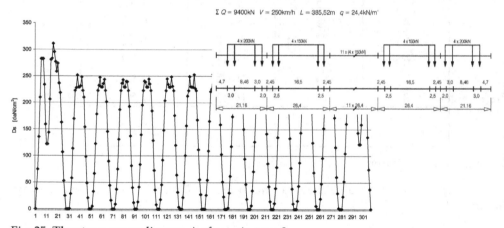

Fig. 25. The stress range diagram Δσ for train type 3

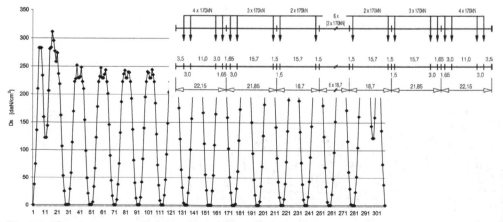

Fig. 26. The stress range diagram Δσ for train type 4

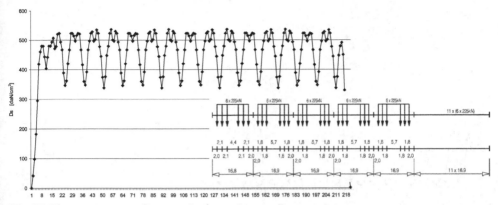

Fig. 27. The stress range diagram $\Delta\sigma$ for train type 5

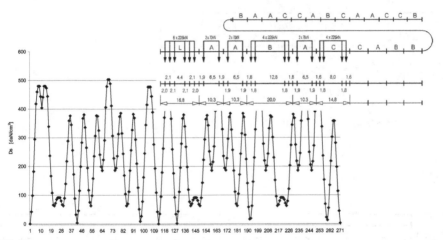

Fig. 28. The stress range diagram $\Delta\sigma$ for train type 6

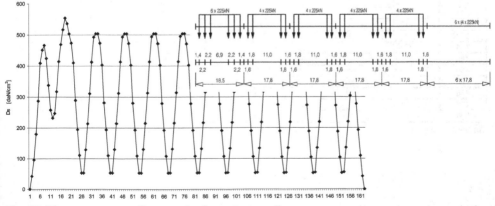

Fig. 29. The stress range diagram $\Delta\sigma$ for train type 7

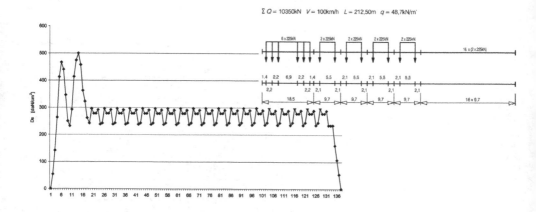

Fig. 30. The stress range diagram $\Delta\sigma$ for train type 8

$\Delta\sigma$	MAIN GIRDER (L/2) mix								
[daN/cm^2]	T1	T2	T3	T4	T5	T6	T7	T8	Total
50	0	0	0	0	105	36	0	108	249
100	0	0	0	0	7	12	0	0	19
150	12	12	10	0	0	0	0	0	34
175	0	108	0	0	0	0	0	0	108
200	132	0	0	10	105	36	8	0	291
250	12	0	65	60	0	0	0	12	149
275	0	0	10	10	0	0	0	0	20
300	12	12	10	0	0	60	0	6	100
350	0	12	0	0	0	60	8	0	80
400	0	0	0	0	0	36	0	0	36
450	0	0	0	0	0	0	88	6	94
475	12	0	0	0	0	0	0	0	12
500	0	0	0	0	14	72	8	0	94
$\Delta\sigma_e$ [daN/cm^2] =							381.36	Total/day	1037

Table 5. Classification of the stress range intervals

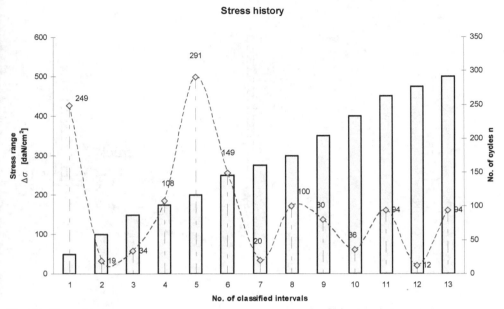

Fig. 31. Stress history

The crack case assumed for these structures (riveted bridges) is illustrated in figures 32; it is a trough thickness defect.

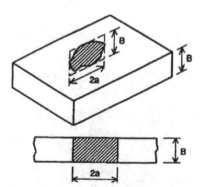

Fig. 32. Theoretical crack models

In order to determine the remaining service life it is important to know how long it will take the crack to grow from the minimum detectable size to the critical value.

Two cases were studied:

- through thickness flaw with initial size $a_0 = 2.0$ mm which is undetectable because it is situated under rivet head in the web steel plate, and
- through thickness flaw with initial size $a_0 = 2.0$ mm which is also undetectable during a visual inspection appearing in the corner (lower flange).

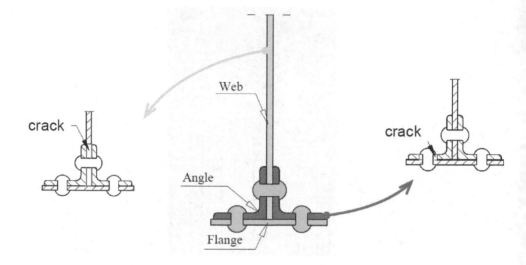

Fig. 33. The two studied defect cases

In the first case the remaining fatigue life is estimated at a value of 8.31 years (corresponding to a number of cycles of 1 566 000). In the second case the administrator must be acquainted with the fact, that propagation of the crack to the critical value will occur in 8.89 years (N=1 676 000 cycles).

In this case the safe inspection intervals, calculated with formula 8, are situated between 1.2 and 2.0 years.

Finally due to the high corrosion level (Figure 34) and also based on the fatigue assessment, the superstructure was proposed for the replacement.

Fig. 34. Corrosion attack details

Also all the other plate girder bridges were verified. They are presented in the next table.

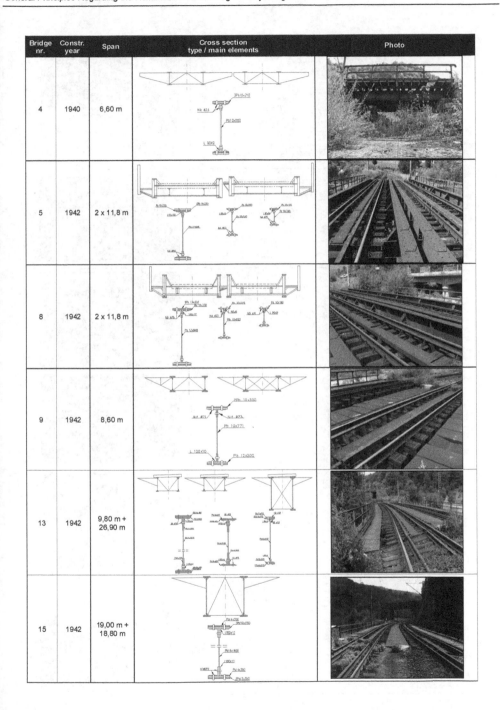

Bridge nr.	Constr. year	Span	Cross section type / main elements	Photo
4	1940	6,60 m		
5	1942	2 x 11,8 m		
8	1942	2 x 11,8 m		
9	1942	8,60 m		
13	1942	9,80 m + 26,90 m		
15	1942	19,00 m + 18,80 m		

Table 6. Analyzed plate girder bridges

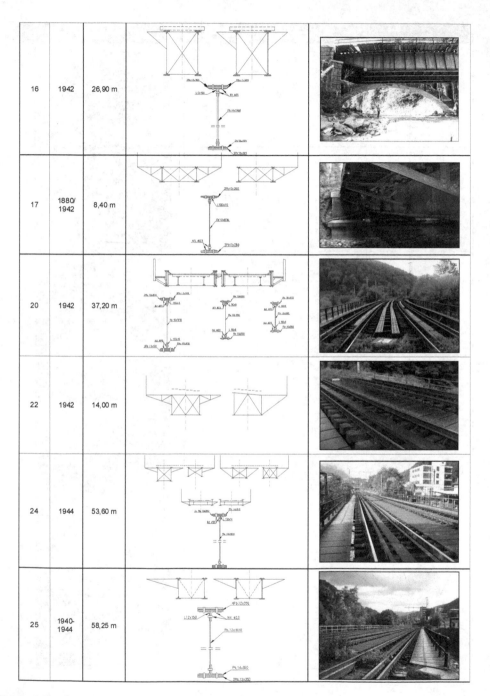

Table 6. Analyzed plate girder bridges - continuation

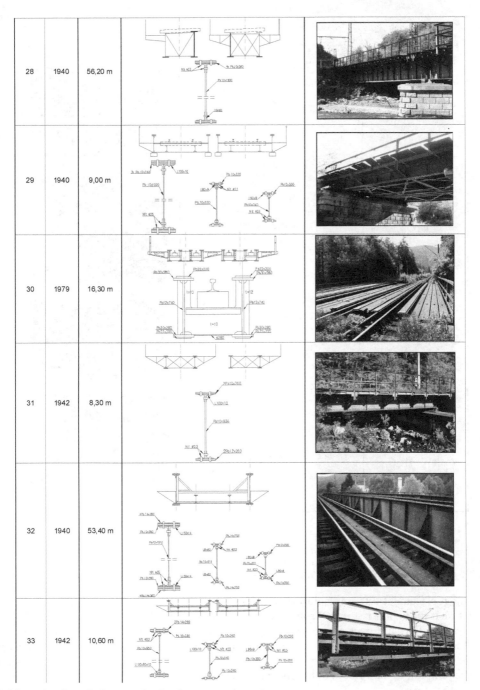

Table 6. Analyzed plate girder bridges - continuation

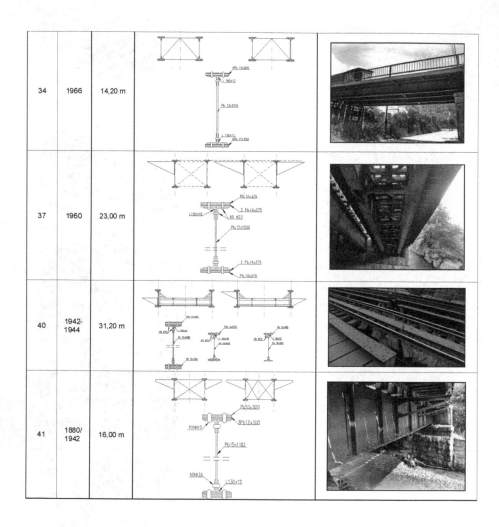

34	1966	14,20 m		
37	1960	23,00 m		
40	1942-1944	31,20 m		
41	1880/1942	16,00 m		

Table 6. Analyzed plate girder bridges - continuation

For the replacement different solutions can be proposed. A modern economical and robust structure is the Schmitt Stumpf Frühauf solution (VFT-WIB®), [12] – which has a high degree of prefabrication and goes along with a significant improvement of working conditions, as weather-independent working; also improved environment conditions for the workers while erecting formwork, placing re-bars and casting concrete is guaranteed. Furthermore a smaller amount of man-hours outdoors at the construction site is needed.

The constructive depth was optimized by a computer specialized program (Figure 35).

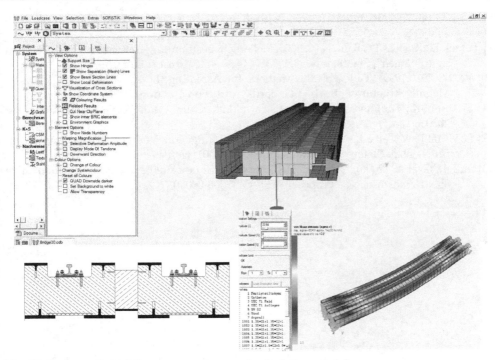

Fig. 35. Cross section of the new structure and calculation model

6. Conclusion

The progress recorded in the last decades allows on one part an accurate evaluation of the remaining safety of the structures and on the others part the proposal of new efficient, economical solutions with the high level sustainability.

7. References

[1] * * * Code UIC 778-2R; Recommandations pour la détermination de la capacité portante des structures métalliques existantes; Union Internationale des Chemins de fer, Paris, 1986.

[2] **, DS 805 „Bestehende Eisenbahnbrücken. Bewertung der Tragsicherheit und konstruktive Hinweise, August 2002.

[3] **, SR 1911-98, „Poduri metalice de cale ferată. Prescripții de proiectare", Institutul Român de Standardizare, Bucuresti, 1998.

[4] Jackson, P. "Is bridge assessment losing its credibility?" The Structural Engineer, Volume 79 / No 9, May 2001.

[5] Xie, M., Bessant G.T., Chapman, J.C., Hobbs, R.E., "Fatigue of riveted bridge girders" The Structural Engineer, Volume 79 / No 9, May 2001.

[6] * * *, "Maintenance of steel bridges", Romanian Pre-standard, Bucharest – 2000.

[7] Petzek, E., „Safety in Operation and Rehabilitation of Steel Bridges", Doctoral Thesis, UP Timișoara, 2004.

[8] ** I-AM 08/2002. Richtlinie für die Beurteilung von genieteten Eisenbahnbrücken, SBB CFF FFS.

[9] Petzek E., Kosteas D., Bancila R., 2005. Sicherheitsbestimmung bestehender Stahlbrücken in Rumänien", Stahlbau Nr. 8, 9, ISSN 0038-9145, Ernst & Sohn.

[10] ***, BS 7910:1999, „Guide on the Methods for Assessing the Acceptability of Flaws in Metallic Structures", British Standards Institution, London, 1999.

[11] Eriksson, K., Toughness requirements for old structural steel, IABSE Report Congress, 2000.

[12] SCHMITT, V., et alt.: VFT-Bauweise, Entwicklung von Verbundfertigteilträgern im Brückenbau, Beton- und Stahlbetonbau 96, 2001, Heft 4.

[13] SEIDL, G. et al. (2006), "Prefabricated Enduring Composite Beams based on Innovative Shear Transmission – Proposal RFSR-CT-2006-00030 ".

## 4

Gaming Simulations for Railways: Lessons Learned from Modeling Six Games for the Dutch Infrastructure Management

Sebastiaan Meijer
1Delft University of Technology
2Royal Institute of Technology, Stockholm
1The Netherlands
2Sweden

1. Introduction

The Dutch railway system is a highly complex and heavily utilized network (Goverde, 2005; CBS, 2009). Worldwide it is one of the most densely driven networks, yet its capacity has to increase further. Improvements in the domain of capacity management and traffic control are increasingly difficult to implement because of the large interconnectedness of all processes. The de-bundling of rail infra management (ProRail) and train services (predominantly NS, and some smaller regional lines by Syntus, Veolia, Arriva a.o. plus freight train operators) has created an operational process in which multiple offices and platform/line operations need to synchronize to control the daily train flow.

ProRail, the Dutch railway infrastructure manager, has stated a goal to increase the capacity by 50% as a challenge till the year 2020. This cannot be done the 'old way' through increased amounts of physical infrastructure, as both money and geographical space are insufficient. Furthermore, the complexity and interconnectedness of the network is yet at such a level that more of this will lead to less resilience and becoming (even more) prone to disturbances. Because of the 50% growth challenge till the year 2020, new and smarter ways of managing capacity and traffic are key for the success of the Dutch rail infrastructure for society. The ProRail organization has taken up gaming simulation as a key method to improve the innovation process (Meijer, forthcoming).

Unique for gaming simulation is the highly detailed simulation of both technical and process variables of rail infrastructures and the decision and communication function of real people in their real roles. The method does not assume models of decision-making but draws upon the real-world knowledge of professionals in the operation. Over the course of the projects that ran in 2009, 2010 and 2011, the specific setting of the ProRail organization proved to be both a complex and fruitful environment for gaming simulation. The complexity was found in the large number of stakeholders both in and outside the organization and in the interconnectedness of every aspect of train traffic control on the performance of passenger and freight train service providers.

In the year 2009, the gaming group of Delft University of Technology was asked to facilitate three projects using gaming simulation methodology. These projects ran so successful that the organization asked the Delft researchers to identify where in the organization large-scale implementation of gaming simulation methodology would be most promising. Based upon a series of interviews through the organization, ProRail and TU Delft jointly formulated a four-year research and implementation proposal that is now in operation. The first gaming sessions in this new collaboration have been held and results have led to methodological lessons-learned on how to model. This chapter reports on three modeling issues crucial to gaming simulation for railway and similar systems. How to abstract from the nitty-gritty details while keeping the simulation real and valid enough for real-world operators to participate and do their job is the focus of this chapter.

2. Problem description

Innovation in the Dutch railways is on one hand much needed, while on the other hand very complex to achieve. The 1995 politically instigated de-bundling of rail infra management (ProRail) and train services (predominantly NS, and some smaller regional lines by Syntens, Veolia, a.o.) has created an operational process in which multiple offices and platform/line operations need to synchronize to control the daily train flow. The increasing importance of rail services for individual provinces in the Netherlands has led to multi-party tendering (Van de Velde et al, 2008). In this complex multi-actor and multi-level environment the strategic safeguarding of public values in managing operations proofs often impossible (Steenhuisen et al, 2009). The combination of these events and trends leads to a challenge to innovate on two aspects, being quality in operations and ways to increase the capacity.

2.1 Quality in operations – Robustness and resilience

Over the past decade, the railways in The Netherlands have received major criticism for the quality of its operations. From a policy perspective this has led to performance contracts for both the main train service operator (NS) and the publicly owned infrastructure manager ProRail (Van de Velde et al, 2009). Over the past decade the performance has seen improvements on the critical performance indicators, but still it is not regarded to be a high quality service due to many small delays, overly crowded trains and non- or mal-informed passengers. The rail system often suffers from small defects, leading to bigger delays when the problems spread like an oil spill over the regions and lines. If we define robustness as the degree to which a system is capable to withstand problems within the limits of the designed system, then the robustness of the railways is questionable.

A lower score on robustness would not have been so detrimental is the railways were more resilient. Hollnagel et al (2006) define resilience as the ability of a system or an organization to react to and recover from disturbances at an early stage, with minimal effect on the dynamic stability. The challenges to system safety come from instability, and resilience engineering is an expression of the methods and principles that prevent this from taking place. Furthermore the recent years have shown that snow, storms, national festivities and other outliers in the situation for which the system is not specifically designed cause total or at best partial collapse of the national system, as soon as small problems start to occur. This has led to Parliamentary Investigation (Rekenkamer, 2011). According to Hale and Heijer

(2006), railways, from their assessment of safety operations at the Dutch Railways, would seem to be examples of poor, or at best mixed, resilience, which can, however, still achieve high levels of safety, at least in certain areas of their operations. Hence safety is achieved by sacrificing goals, traffic volume and punctuality. The system does not achieve all its goals simultaneously and flexibly and is not resilient.

2.2 Capacity increases

The Dutch railway sector will face a massive growth of transport demand in the forthcoming decade. This growth is both expected in passenger and in freight transport. Currently, the Dutch railway network is one of the most densely used networks in the world, approaching its maximum capacity given the current infrastructure and control mechanisms. The projected increase in demand requires a step-change in both the physical and control aspects of the railways. ProRail formulated an ambitious program, called 'Room on the Railways" (Ruimte op de Rails, in Dutch) to increase the number of trains on the network by 50% before the year 2020. One of the major components of this program is the plan for high-frequency passenger trains on the major corridors. Currently there are (on average) 4 intercity, 2 to 4 local and 1 or 2 freight trains per hour on the major corridors. This should increase to 6 intercity, 6 local and 2 freight trains before 2013. This new frequency of trains is often called 'untimetabled travelling" as the passenger can just go to a station without checking departure times: the next train will be there soon. The official title of the schedule is High Frequency Train Transport.

The projected increase of capacity cannot be achieved by building new infrastructure alone: the costs for the complete program would be around 9 billion euro, and the time for procedures and construction would frustrate the transport demand for years. ProRail has taken up the challenge to achieve the goals with only half of this budget by combining strategic choices for new infrastructure with new control and management solutions.

3. Gaming simulation for process innovations

Gaming simulation, here defined as 'simulating a system through gaming methods' is one of the terms in a loosely demarcated field of interactive participatory activities, aiming to involve participants, who may be the real stakeholders, in an activity. Other terms used are simulation game, policy exercise and serious gaming. The word gaming will be used here as the short term for gaming simulation. Different authors have different preferences, but in in general the terms depend on the intended use of the method.

Game theory and gaming simulation are often intertwined. Game theory is the mathematical approach of analyzing calculated circumstances where a person's success is based upon the choices of others. In games, these situations often occur, and therefore is game theory a popular method of modeling artificial intelligence in games. This chapter does not use game theory per se, however a more prominent and explicit use is foreseen in future games that incorporate automated agents.

Given the number of gaming titles and scientific publications, the use of gaming methods for learning is the most popular by far, typically occupying 'serious gaming' and 'simulation game' for usually computer-supported games that place the player in a simulated world

(Bekebrede and Mayer, 2005; De Freitas and Martin, 2006; Kriz and Hense, 2003). Learning about innovation in games is a popular topic for MBA-style versions, typically related to markets and supply chains (Meijer et al, 2006; Meijer, 2009). Learning and communicating complex issues are in this stream highly related. An important aspect for ProRail is the opportunity to communicate ideas. While a slideshow can communicate a message, a gaming simulation enables you to experience it for yourself (Bekebrede, 2010). The aspects about which it is sometimes difficult to communicate at present include: the impact of new timetables (on all categories of employees), the need for precision in carrying out tasks (employees), the influence of disruptions on the network as a whole (general public) and to experience the key aspects of traffic control / capacity management (general public). At present, visualizations of train flow models such as FRISO and SIMONE (Middelkoop and Loeve, 2006) are available, but it is not possible to experience these aspects by sitting at the controls. The opportunity for communication gives employees the chance to play a role that they do not have in reality. This can help clarify different points of view.

In the world of policymaking, there is half a century of history in using gaming as an intervention to bring together policy makers and other stakeholders in participatory events. Games provide a way to collectively decide firstly on the system boundaries and secondly on the dynamics of the system that will be played. Then, policies can be formulated in this simulated environment (Duke, 1974; Duke and Geurts, 2004; Mayer, 2010). This approach relies on Duke and Geurts' (2004) 5-C's of gaming simulation for improving policy making, namely by understanding the Complexity, enhancing Creativity, enabling Communication, reaching Consensus and Commitment to action. Within ProRail this role of gaming simulation is particularly relevant for management questions.

Increasingly popular is the possibility to try out the effect of policies on a simulated system, and see whether innovation in roles, rules, objectives and constraints can be made. This approach, although very relevant for policy-making, is actually a third use of gaming, for testing hypotheses (Peters et al, 1998). This application is less common and puts great emphasis on the verification and validation of the gaming simulation (Klabbers, 2003, 2006; Noy et al, 2006; Meijer, 2009). For innovation at ProRail, this use is at the core of the reasoning behind choosing gaming simulation as a new method in reducing uncertainty in more complex, system level changes.

A fourth use that is emerging is linked to the gamification of society (Hiltbrand and Burke, 2011). Innovation can take place through game play if the incentives are such that the crowd can generate and implement their ideas in a system. Few scientific literature on this exists as of yet, but examples are UK innovation in pensions (Gartner, 2011), crowd sourcing of ideas in an insurance company (Bekebrede and Meijer, Forthcoming)

4. Modeling challenges in gaming simulation for railways

In the world of gaming simulation several design guides and principles exist on how to capture real world problems in a gaming simulation. In the field of policy making the most important method is the one that Duke and Geurts (2004) describe, where for learning, sense-making and related issues the more recent work of Harteveld (2011) is gaining footage among some others. However, the problem is that these methods are so generic that the specific issues for technical domains like the railways have to be addressed specifically for that domain. A second issue, especially with the more policy-oriented approaches and the popularity for learning in higher education is the focus on participants with a relatively large capability in thinking abstract, as policy makers and students tend to have more of this skill than the average operator. Peters et al (1998) describe the process from real world to simulated game world as a process of abstraction and reduction. The big question is how far can you abstract and reduce from reality before operators loose their grip on the simulated reality?

The operational skill training is recently getting more and more addressed in the gaming literature. Druckman (1994) proved already the need for more 'fidelity' (that could be translated as 'detailed realism') when training less abstract skills. Applications for operational skill training is getting common in the domains of image-based medical procedures (like laparoscopy, gastrointestinal flexible endoscopy, image-guided neurosurgery, and endovascular surgery) (Gaba, 2001; Botden et al, 2008; Hamdorf and Hall, 2000), aviation (Proctor et al, 2007), and safety training for dike inspection (Harteveld, 2011) and the oil and gas industry (Meijer and Poelman, 2011). Each of these domains finds a solution in 3D-based computer games that model an environment, either geographical or the organs in a body, through which the player has to navigate and perform a coherent set of actions. There is an overlap between the fields of virtual reality, simulation and gaming here.

Involving operators in games for policymaking or for testing hypotheses is almost undocumented, with some notable exceptions like the work at CIRAD and Cemagref (Barreteau, 2003). Traditionally the questions in policymaking and the hypotheses tested are at a higher level of abstraction. In Meijer (2009), the author argued that involving the real operators in a gaming simulation has the benefit of avoiding models and assumptions about their behavior, and thus can increase the validity of the behavior of the entire socio-technical system simulated. This has been proven in the domain of supply chain management research, studying the organization of transactions.

In our work we focus on the behavior of the people in the daily operations in railway systems, with a focal point at the train traffic controllers. Within the scope of the infrastructure management ProRail, their behavior has the most direct influence on the robustness and resilience of the network. To base decisions upon their behavior in gaming simulations it is essential to consider the validity of this behavior.

The most common critique for behavior observed in a session is "it is only a game.....". In the literal meaning the statement is true. A gaming simulation is a model of reality, and the roles, rules, objectives and constraints are necessarily different from the real world. The insinuation of the statement is, however, that behavior observed in a session is unlike behavior in the real world and is no valid representation of real-world behavior. Peters et al (1998) discuss the validity of games (gaming simulation) based upon the work of Raser (1969) who defined validity of models in the following way: "A model can be said to be valid to the extent that investigation of that model provides the same outcomes as would investigation in the reference system." Raser (1969) suggests four aspects of validity that apply to gaming simulation:

- Psychological reality: To what degree does the gaming simulation provide an environment that seems realistic to the participants?
- Structural validity: To what degree is the structure of the gaming simulation (the theory and assumptions on which it is built) isomorphic to that of the reference system?

- Process validity: To what degree are the processes observed in the gaming simulation isomorphic to those observed in the reference system?
- Predictive validity: To what degree can the gaming simulation produce outcomes of the historical or future reference system?

The psychological reality demands that sessions are conducted in such a way that participants are emotionally involved and really play their role. The situation of the session in the life of the participants, the consequences of participation or non-participation and the location and atmosphere of a session and its moderation are important factors. This requires craftsmanship of the game leader that is hard to operationalize in a scientific context. Various authors have made attempts at determining the quality of conducting sessions. Kriz and Hense (2006) offer an elaborate and theory-based evaluation methodology, that according to Klabbers (2008) does a good job in (temporarily) bridging the gap between analytical and design sciences. Kriz and Hense's approach is an adapted version of the theory-based evaluation method by Reynolds (1998). They distinguish between concept, design and application that can be evaluated.

The psychological reality and process and structural validity of Raser (1969) come together in the concept of situation awareness (Endsley, 2004) for operators. Operators should get involved psychologically when they can recognize sufficient components of the processes and structures they are used to in their real work. In the medical world this has led to consensus guidelines for validation of virtual reality surgical simulators (Carter et al, 2005), but in railways this work is only done for train drivers (Hamilton and Clarke, 2005)

When we take the concept of situation awareness as the central concept for considering the validity of railway operator behavior in gaming simulations, the list of items in the situation awareness still becomes vast. The modeling issue could include nearly any technical aspect of the railways, interface and representation items as well as the cognitive state of operators during their normal workdays. This chapter focuses on three important issues, posed as how-to questions:

1. How to immerse train traffic controllers in a gaming simulation?
   Immersion is one of the important indicators of presence and therefore psychological reality in simulated environments (Witmer and Singer, 1998). Therefore an indication of how to model a game so that railway operators get immersed in a first important step towards validity.
2. How to model time?
   Where real-world train flow is a continuous time process, this does not necessarily translate one-to-one to a gaming simulation, as the research question may ask for another solution than continuous, like step-wise, round-based, or asynchronous time.
3. How to present operational data?
   In the real operations the data flow to operators is bundled in machine interfaces or is fairly constant as time tables typically change only once a year, and infrastructure doesn't change fast either. In a game the standard tools may not be available and timetables and infrastructure may be new to the participants. How to present the information so that operators can still use it?

These questions will be answered in the remainder of this chapter. The next section discusses six projects from which the experiences are gained, then Section 6 translates this into lessons learned.

5. Modeling gaming simulations for rail innovation projects

As of 2011, the gaming collaboration resulted in six different gaming simulations, specifically built for innovation projects within ProRail. Each of these projects used gaming simulation to investigate various solution strategies with the aim of increasing capacity utilization, resilience and robustness on the rail network. The initial pilot project covered three projects while the long-term collaboration yielded three so far.

From the launch of the initial project, ProRail formulated three preliminary cases to study using gaming simulation. TU Delft was to developed unique approaches for each of these cases, after which the initial success of gaming simulation for the Dutch Railways would be re-evaluated. The cases differed in nature. The first was about the potential value of market mechanisms for management of demand of cargo capacity. This game could be seen as a management game on the tactical level. The second case was about studying a control concept for high-frequency train transport at the Bijlmer junction. This game was at the operational level of train dispatching and network control. The third case was about the opening regimes of the bridge over the river Vecht. This game was purely about train dispatching at the operational level.

During the course of these three cases, the success became very apparent to the senior management involved at ProRail. This led to an Intermezzo phase after the third game to reflect upon the results so far and to identify the value from interviews with ProRail internal stakeholder held by Delft researchers. The launch of a large four-year project was marked by a kick-off case that convinced the last skeptics. In the following sub-sections each of the cases and the intermezzo phase are described.

5.1 Rail cargo market game (RCM)

The first and kick-off subproject called Goederenmarktplaats (Freight Market) introduced ProRail to a paper-based and partly computer-supported game with a high degree of abstraction. This game type was referred to as a management game, due to the focus on more abstract policy-related aspects. Most of the participants were managers, with one session including a small number of network controllers.

Table 1 lists the core description of this game, more information can be found in Meijer et al (2009).

The game sessions delivered results timely, and in a positive and active manner. This game is still referred to two years later in the organization. Important to note for the introduction of gaming is that this project happened to have many people on board in senior staff functions from two different divisions (Traffic Control and Capacity Management) who appeared to be key people in later problems that called for gaming simulation methodology. The foundation in terms of exposure to key personnel therefore couldn't be better.

The research team then conducted interviews within ProRail to evaluate the pilot project and identify the opportunities it presented. In these interviews, the management game was repeatedly described positively. However, this generated few new ideas as regards applicability. Many of the issues encountered within the ProRail organization are operational and thus call for less abstract forms of gaming simulation.

Core aspect	Description
Purpose	Studying the potential value of various market mechanisms for better capacity allocation of cargo paths.
Roles	Clients with demand for transport, Rail Cargo Transporters, Passenger Transport, Rail Capacity Planning, Rail Asset Management
# of players	15 – 25 depending on step
Own/real/fictitious role	Real role, but selected for knowledge for instance from previous job position.
Scenarios	3 – 4 scenarios per session. First scenarios that explored the more fundamental market mechanisms. Then scenarios to validate the successful configurations.
Intervention range	Facilitator could start and stop the scenario and dissolve disputes only on the process steps.
Simulated world	Stylized train path market, stylized transport demand
Immersion	Fast, once roles were clear and adopted. Lively play including some conflicts. Capacity planners in second session had issues getting insight in their track system.
Time model	Continuous
Data presentation	Simplified to stylized network, simplified timetable and simple contracts. Big jump between session 2 and 3 when replacing capacity planners with computer reservation system that was similar conceptually.
# of sessions	3 subsequent games each with 1 session during 1 full day.
Type of data generated	Quantitative and qualitative, testing hypotheses about mechanisms that are assumed to have a certain effect on capacity allocation.
Consequences	Policy formulated but put out of scope for 2010/2011, possible application in 2012. Politically very sensitive.

Table 1. Core description of Rail Cargo Market Game

With respect to the three modeling issues we learnt that it is very hard to have operational level people in an abstract simulation when they have to work on infrastructure and timetabling they do not recognize. The usual flexibility that is commonly found in gaming with groups of higher education is not working here. To overcome this, we automated some of their tasks in the game into a computer model for train path reservations. This worked flawless for the more management-like question of this game.

5.2 Bijlmer junction game

This subproject introduced ProRail to a computer-based gaming simulation developed on ProRail's own MATRICS simulator (Van Luipen and Meijer, 2009). This simulation pushed the envelope in terms of utilizing the technical specifications of MATRICS. This type of game was described as a multi-player process simulation due to its detailed reflection of real-life operational processes. The participants play a pre-defined role that is 100% identical to their job description, to carry out their real-life duties in a simulated game environment. Table 2 lists the core description of this game. For a full description we refer to Meijer et al (2009).

Core aspect	Description
Purpose	Testing and validating a control concept for high frequency train transport.
Roles	Train driver (2), Train traffic controller (3), Network controller (5)
# of players	10 plus 2 facilitators and 2 experts.
Own/real/fictitious role	Own role, participant selected by their team leaders
Scenarios	3 Scenarios, gradually testing more complexity.
Intervention range	Facilitators could start, stop and pause scenarios and interfere with train driver behavior.
Simulated world	Detailed infrastructure between Amsterdam and Utrecht, detailed timetable.
Immersion	Very fast and deep for train drivers and network controllers. Difficult for train traffic controllers; see discussion in Meijer et al (2009).
Time model	Continuous
Data presentation	Highly detailed through computer interface. Interface different from real-world abstraction.
# of sessions	1 full day session
Type of data generated	Quantitative (failed) and qualitative.
Consequences	Data generated in the game yielded insights in key materials and resources needed for implementation of the control concept, and high-frequency planning in general.

Table 2. Core description of Bijlmer Junction Game

ProRail had assigned a project team to come to new control and steering procedures that suite the future reality of high-frequency passenger trains. The challenge of this project team was to come up with new concepts that would both be supported by train traffic controllers and network controllers, and would yield a stable, controllable control and routing operation when put into place. The question was raised: how to test new control and steering concepts when there is no option to test in real life? The Bijlmer Junction Game was targeted at this. In the game the interaction of train drivers, traffic controllers and network controllers was crucial, as studied earlier by Albrecht (2009).

The gaming simulation session yielded insights in key materials and resources needed for implementation of the control concept, and high frequency planning in general. The importance of buffer areas with sufficient space to side-track a train without disturbing other services, platforms asides the entire train for passenger exit, and alternative departure options for all passengers within reasonable time is a clear outcome for ProRail. Furthermore, train traffic controllers do not yet seem to realize what the projected high-frequency planning will mean in practice for their tracks.

As described in Meijer et al (2009), this game was not a break-through success. We learned that involving the operational people in the organization in a game that modeled the infrastructure and timetabling as detailed as they are used to, requires interfaces that connect to the situation awareness capabilities of these operators. Simple said: even though we checked our approach upfront with the operators, they were not able to do what they though were capable of due to different visualization. Luckily, the debriefing and discussions still yielded sufficient data of sufficient quality for ProRail to be able to contribute to the problem solving. For the gaming team, this experience led to the development of the following game.

5.3 Railway bridge game

The subproject Railway Bridge Game (for a bridge over the river Vecht) introduced ProRail to the process management game, a computer-based gaming simulation for which new software was developed. Over the course of one week, various train traffic controllers played this game in a single-player environment using a series of scenarios. The type of game was described as a single-player process simulation. Table 3 gives the core description of this game. More information can be found in Kortmann and Sehic (2010).

Core aspect	Description
Purpose	Studying a new regime for bridge openings on the busy Amsterdam – Amersfoort corridor.
Roles	Train traffic controller. Bridge operator (simulated)
# of players	1
Own/real/fictitious role	Own role.
Scenarios	5, each subsequent day the same train traffic controllers played one scenario of increasing complexity
Intervention range	Facilitator played other roles
Simulated world	Detailed infrastructure, detailed timetable
Immersion	Good to very good. More experienced train traffic controllers had more hesitancy towards the computer system, but once used to it scored better with more situation awareness.
Time model	Continuous
Data presentation	Detailed through near-familiar computer interface.
# of sessions	1 session, full week
Type of data generated	Mainly quantitative (measured actions and train throughput, questionnaires) and qualitative from interviews
Consequences	None as of 2011, new game with improved interfacing planned for winter 2012 testing more details.

Table 3. Core description of Railway Bridge Game

The Railway Bridge Game was positively received. It learned that the drawbacks of the interface problem signaled in the Bijlmer Junction Game could be overcome by making special gaming modules. In these modules the representation of the infrastructure and the control options can be made closer to the real world systems. Given the differences between experienced and less experienced controllers we conclude that more resemblance is better for immediate immersion, but not necessarily related to the quality of the decision once a certain threshold of realism is reached.

Playing the game showed its potential to help solve the bottleneck of the Vecht Bridge on the OV-SAAL rail corridor. Under increasing loads of timetabling the experienced operators scored significantly better than operators in training. Under light loads this was the other way around, showing the difficulty of experienced people to overcome differences in the user interface, but as soon as craftsmanship was required to minimize delays the experience helped keeping control.

Both single-player and multi-player gaming simulation were readily welcomed by almost all of the stakeholders in the organization as a valuable new resource for ProRail as an organization. The aspect of the multi-player gaming simulation that prompted a particularly positive response was the opportunity to test the feasibility of timetables, control concepts and exceptional situations in a setting that includes several layers of management and/or control areas. The aspect of the single-player gaming simulation that prompted a particularly positive response was the opportunity to train and practice in relation to exceptional situations and future timetables and infrastructures in an offline setting, using simulated trains.

5.4 ETMET 2010

One of the two strategic innovation trajectories to come to the desired capacity increase is the program to come to a metro-like timetable on the major corridors. On the Amsterdam – Eindhoven corridor this program is titled 'Every Ten Minutes A Train' (Elke Tien Minuten Een Trein – in Dutch), shortly ETMET. In the fall of 2010, the largest train operator National Railways (NS) and ProRail tested this concept for a full month in the real operation. This program required substantial preparation, and gaming simulation was selected through the senior staff involved in earlier games to answer questions about two ways of handling a major disruption under the new timetable. This resulted in the ETMET 2010 Game, described in Table 4.

In the ETMET 2010 Game we simulated the train flow and all processes and interactions in the train control, personnel and rolling stock processes. The wish was to have the train traffic controllers working on gaming modules similar to the one in the Railway Bridge Game. Soon during the development we found out that the underlying rail traffic simulators available did not support the required actions of turning around, skipping a service or renumbering rolling stock to different train services. Therefore the decision was made to create a complete manual, analog simulator, observed with cameras overhead the infrastructure maps, distributing views similar to the regular computer visualizations to three rooms with operators.

The session delivered the data required to answer the question on the differences between two methods of handling a major disruption. The project management assumed the new method to be beneficial for resilience, however they proved wrong. The new method essentially provided a pre-defined pattern for guiding trains over a double track where one track is blocked. The network and service controller had to makes their choices out of the set of trains currently running on the tracks, approaching the blocked track. Remaining trains have to be cancelled or coupled. This was assumed to be a better solution than the old solution in which there is a separate document for every possible interaction between two trains. It appeared however, that the choices for assigning trains to the pattern were impossible to make, given the interactions that all the trains available have with other parts of the system. While working on a solution the situation changed too fast to make a single decision in time, while overseeing all of the complexity.

Core aspect	Description
Purpose	Testing the differences between two mechanisms of handling a major disruption under High Frequency Transport scheduling
Roles	Train traffic controllers, Passenger information, Driver rescheduling, Rolling Stock rescheduling, Platform coordinator, Decentralized network controller Network controller, Service controller.
# of players	14 in role, 9 in support roles in analog simulator center, 6 observers, 1 host, 1 game leader
Own/real/fictitious role	Own roles, invited on personal title however with support of management.
Scenarios	2 scenarios: first the 'old' way and then a new mechanism
Intervention range	Facilitators could start, stop and pause the scenarios.
Simulated world	Detailed infrastructure Utrecht - Geldermalsen, detailed high-frequency timetabling, essentials of communication lines between different offices involved. Stylized passenger flow.
Immersion	Full immersion in a few minutes. Conflicts arose, leading to a time-out by the game leader to settle the issues and go back to a state all could agree on. Extremely involved and lively game play.
Time model	Continuous
Data presentation	Infrastructure representation in familiar schematics, detailed timetabling on paper, time and delays through simple interfaces.
# of sessions	1 session, full day
Type of data generated	Quantitative and qualitative, testing hypotheses about differences between 2 mechanisms.
Consequences	Proposed solution abandoned based on data generated in the gaming session.

Table 4. Core description of ETMET 2010 Game

In this game all data was completely detailed available to all participants, on paper, and mostly in a format they recognize, using all real-world abbreviations and notations. Contrary to many games in which the designer abstracts and reduces to a level where it is not about managing large amounts of data, it proves to be very well possible to give operators this data. It even helps to give them situation awareness, as in the debriefing all were confident that their behavior and reactions were similar to what they would to in practice, and all could work with the data supplied.

5.5 NAU

Utrecht Central station is the heart of the Dutch rail network: here come trains from all directions together in a versatile, but consequently complex knot. The complexity and interdependence of the many train movements and other activities makes Utrecht very vulnerable to chain reactions of delays. For large disturbances, history has repeatedly shown that the risk of flooding and even completely crashing the traffic flow is high. Resilience and robustness of the Dutch rail network therefore has to consider Utrecht as a key parameter.

In previous years the timetable has been 'disentangled', meaning that trains are assigned to a corridor and that these corridors are planned to have as little interference with each other on the physical infrastructure as possible. Now, after ProRail disentangled the schedule the aim is to reduce the interference further by matching the control concept to the corridors. For this the NAU (New Action plan Utrecht) program was launched.

Within the NAU program five goals have been identified:

1. ensure that the basic plan remains within the corridors;
2. limiting defects through maintenance / inspection;
3. Limiting deviations from shunting;
4. limit abnormalities in major disruptions;
5. adjust the division of labor in the Traffic Control Post.

Gaming simulation has been selected as a tool to try out the concept first in a simulated environment before it will be brought to the control post. During the game the effect of the new concept on goals 1,4 and 5 had to be researched. Table 5 describes the game details.

The main result of the NAU game was empirically based insight in the fundamental consequences of reducing the number of switches used and corridor control on capacity, resilience and robustness. In the old situation, the capacity reduces rapidly as the disturbance level increases. Due to the many switches, many options remain in heavily disturbed situations to continue driving, requiring a mastery level of the train controller. These options cause 'infection' of problems of one corridor to others. When using corridor control in its strict sense, the process remains more manageable with mild to moderate disturbances. But because the control options are limited to the corridors, there comes a moment in the corridor that all capacity is lost, still not affecting the other corridors. There's a tipping point where the limits for a disturbed corridor become unacceptably high. At this tipping point it can help to deviate from the corridor principle. The ideal situation is to control & isolate the corridor as long as possible and use other parts of the infrastructure only when the critical level of disturbance is reached. In other words: you want to cash the potential of corridor control and avoid potential losses. Where the disturbance level is critical, how often this situation occurs and what specific deviations must be allowed, is still open for further research.

In the NAU game modeling we re-used the infrastructure schematic layout and timetabling information that was so successful in the ETMET game. Yet again this proved to immerse the participants in the simulation within a few minutes, and to make them enact their role perfectly. In this post-game evaluation the participants rated their behavior as highly realistic. The only exceptions to this were the network and service controller who both work at the national level. For them there was no game material to play with apart from information derived from the simulated area. This resulted in less emersion and a bit grumpy atmosphere in which they were mocking about the new concept. For the project this proved functional as their comments in the discussion raised important points for the improvement of the concept, but the game play from them was not optimal. Therefore no direct conclusions could be drawn from the interaction between the national and regional level.

Core aspect	Description
Purpose	Testing the improvements in resilience and robustness when introducing a new control concept for Utrecht Central station.
Roles	Train traffic controllers, Decentralized network controller, Driver rescheduling, Rolling Stock rescheduling, Platform coordinator, Network controller, Service controller.
# of players	9 in role, 4 in support roles for analog simulator, 3 observers, 1 host, 1 game leader
Own/real/fictitious role	Own roles, invited on personal title however with support of management.
Scenarios	2 scenarios: first the 'old' way and then a new mechanism
Intervention range	Facilitators could start, stop and pause the scenarios.
Simulated world	Detailed infrastructure Utrecht Central, detailed current timetabling, face-to-face communication lines between different offices involved. Stylized planning tools
Immersion	Instant and very good for all players, except for network controller and service controller who were less immersed, showing in discussions about other topics.
Time model	Continuous
Data presentation	Infrastructure representation in familiar schematics, detailed timetabling on paper, time and delays through simple interfaces.
# of sessions	1 session, full day
Type of data generated	Quantitative and qualitative, testing hypotheses about improvements with new control concept. Numbers real enough to base decisions on.
Consequences	New control concept embraced, actions defined to counterbalance penalty for major disruptions. Invention of the concept of pre-defined handling scenarios for non-availability of small parts of the infrastructure. Heavy post-game discussions leading to high-level decisions on the project.

Table 5. Core description of NAU Game

5.6 Platform overnight parking (POP)

In the capacity planning process for 2012, two service areas have been declared out of capacity and ProRail is obliged, according to law, to find solutions to solve these capacity shortages. The goal of this game is to determine whether it is possible to orchestrate a 'carousel process' around Hoofddorp so that scarce capacity in the service area can be increased. This means that after servicing and technical controls at the service area the train is then drawn along the (platform) tracks of Hoofddorp station or Middle Hoofddorp.

Core aspect	Description
Purpose	Answering the question how many pieces of rolling stock could be parked along the platforms of stations during the night, given the processes of cleaning and maintenance that have to be performed at service areas. Question asked for two locations: Amsterdam-Watergraafsmeer and Hoofddorp.
Roles	Train traffic controller (2), Foreman of cleaning (2), Train driver, Service area supervisor.
# of players	6 in role, 2 support for illiterate cleaning foreman, 3 observers, 1 host, 1 game leader
Own/real/fictitious role	Own roles, invited on personal title however with support of management.
Scenarios	1 scenario per location (Plus 30 minutes 'warm up' scenario)
Intervention range	Facilitators could start the simulation and determine the time required for each 5 minutes of simulated time.
Simulated world	Detailed infrastructure Amsterdam Central – Amsterdam Watergraafsmeer and Hoofddorp, detailed timetabling for end-of service of trains. Detailed service demand,
Immersion	Immediate for train traffic controllers and service area supervisor. Foremen took some time, being illiterate and not used to any abstraction, but came in role in 15 minutes.
Time model	Step-wise.
Data presentation	Infrastructure representation in familiar schematics, detailed timetabling on paper, time through simple interface, cleaning capacity in simple game objects.
# of sessions	1 session, full day
Type of data generated	Quantitative and qualitative, delivering a range of rolling stock feasible to park
Consequences	Potential yield for Amsterdam-Watergraafsmeer too unsecure given additional complexity of extra night maintenance in the years 2012 – 2013. Solution considered for 2014. For Hoofddorp the yield found in the game was verified in the field and implemented for 2012.

Table 6. Core description of Platform Overnight Parking Game

By 'gaming' these processes it should become clear whether and to what extent the (platform) rail capacity can be used for the preparation of passenger rolling stock. If a carousel process is theoretically feasible, then follow-up actions are defined to carry out a practical test. Table 6 describes the core features of the game.

The game delivered the results requested in time, by which it became the first game in the row of six that not only drew conclusions based upon the mechanisms of the game play, but also on the numbers generated in the game. Halfway the first scenario in the session there was an intervention required because the service area supervisor felt that the game play was 'not realistic'. After a thorough joint review it appear there were two trains left in the wrong location. The game leader corrected this, and from this point all agreed the outcomes were valid and representative for a normal evening with no major disturbances.

In this game the modeling of infrastructure and information followed the infrastructure schematic layout and timetabling information that was previously successful in the ETMET and NAU games. The time was for the first time not continuous but step-based. The reason for this was the long time to simulate (6 hours) during which many moments are trivial, as trains stand still and some cleaning is done. As no more game time than 2 hours per scenario was available, a speed-up was required, but just faster time would not contribute to the game as some time periods need more attention than others. The solution was found in 5-minute time steps in the game time that could take anywhere between 30 seconds to 20 minutes to execute in clock-time. In this way the players (most of them operational practitioners, two of them near illiterate) could keep up with the more abstract representation of their real work. The two foremen of the cleaning teams had most issues getting involved. Their whole task consisted out of taking 1 toothpick out of a wagon once it was cleaned, and they could each take out one stick per 5-minute step. Once they got used to this task they could make choices for priority over the service area easily and got their behavior realistic according to both their own and others judgments.

6. Methodological challenges

In this section the lessons learned on methodological challenges are discussed on two levels. The first subsection answers the three modeling issues raised in Section 4. The second subsection discusses how to guarantee validity in gaming simulations for railways.

6.1 Modeling issues

In the six projects, the modeling issues appeared to have a large interaction. The question how to immerse train traffic controllers in a gaming simulation appears to be largely dependent on the display of information. For train traffic controllers we learned that a detailed representation of the infrastructure is key for their involvement. However, the geographical representation did not work, where the common abstracted versions as used in practice performed flawlessly, both in digital (RBG) and in analog game board format (ETMET, NAU, POP). More abstraction and reduction of complexity of the infrastructure does not work for operators (RCM). For the timetabling and similar information like personnel and rolling stock planning similar mechanisms worked: give the players the information on a detailed level but keep the format they are used to in practice, even tough the medium (computer or paper) is different. The same held true for delays and other process information. Once the delays are presented directly after a train number in the format of +3, +5, + 10 minutes everybody understood it immediately. Once the players could understand the information well they could concentrate on their task, which they automatically did fully immersed.

Regarding the question how to model time we learned that the logical model of continuous time for rail operations works well and puts pressure on the process. In the ETMET game the frustrations over problems under time pressure became so high that the game leader had to intervene, and other games showed real pressure on the players who are so aware of the real-time nature of their real-world process that this can be triggered immediately in the game. Care should be taken to give players sufficient situational awareness without all their real tools available. Based on the experiences with the Railway Bridge Game versus the analog games to expect that continuous time will improve on stress and pressure level when computer models are more easily deployable and integrated in the games. Analog simulators are surprisingly good but require extensive and thus expensive expert support.

6.2 Validation

The sessions usually run only once. Drawing conclusions on just one session puts emphasis on the validity of the behavior observed and decisions made in the simulation. The number of people to validate a full game with is limited in terms of availability (they work in de 24/7 operation) and costs, validation approaches need to be done differently. By modularizing the toolkit of gaming into sub-models and software components, validation of the components can be done outside of the final game sessions in analogy with the recent insights in multi-agent simulation of social systems (Gilbert, 2011). Work on the validation requires deeper understanding of train traffic control and train driver behavior. This encompasses the knowledge base in the organization. Work on this gives methodological challenges that go beyond the literature on gaming methodology (Peters et al, 1998).

In the railway gaming simulation described above (but for the RCG) processes of self-validation were used to overcome the validation issue for now. During every session signs of discomfort of the players and comments on 'how real' something was were constantly monitored and discussed openly even if this led to time-outs or moments of difficult discussion during the game play. The game leader always stated that everything to make the session better would be welcome at any time. In the debriefing the explicit questions were asked: which part is realistic and which part would be different in the real world, and why? This gave often very valuable information, even when in case of the NAU game when the network and service controller were not very involved in the game play, but could comment and criticize the validity very well.

By ensuring immersive game play and having the self-validation during the games the Raser categories of psychological reality and process validity are addressed to an extent that is satisfying for the organization. The structural validity is a design issue and is difficult to improve when using analog simulators. You simply cannot model the exact train flow and safety and interlocking systems in an analog way. Computer simulators have a lot to offer here in interaction with the players during a game session. This is future work for integration. The predictive validity is currently under review as the project follows all game projects longitudinally to determine the extent to which the conclusions based upon game sessions hold true in the real operation. Future work will report on outcomes.

7. Conclusions

The series of six projects shows the purpose and usefulness of building gaming simulations to help the Dutch railway infrastructure manager ProRail innovate on its core processes. Over the projects methodological lessons on involving operators as game participants have been drawn, as well as for the abstraction and reduction of information and the modeling of time. These modeling challenges appear to be highly interrelated. The lessons learned show the need, contrary to the traditional modeling approaches in gaming simulation, for very little abstraction and reduction in modeling the game where it concerns items that the operators have to play with. The model for less operational aspects can be more abstract, in line with literature on the need for fidelity for learning in games. While this finding may be not surprising to experienced game developers, the value of using abstractions that are used in the real world for the game is new.

As the game projects reported in this chapter are not for learning but for testing of designs and hypotheses, the findings on how to make real operators show valid behavior in a game session contributes to the small but growing field of gaming simulation for testing. For the categories psychological reality and process validity the current approach has found ways to address given the limited time and capacity available for traditional validation. For structural validity and predictive validity future work is defined.

The sequence of gaming simulations led to a successful introduction in the ProRail organization of the gaming method. Full support has led to a four-year partnership between academics and the operation to make gaming suited for ProRail and ProRail suited for gaming. Once this project has been carried out, ProRail will have at its disposal a gaming suite that connects with existing rail traffic simulators. The gaming suite will make it possible to configure a game simulation session without the need to call in outside expertise by selecting timetables, locations, actors, duration and measurement variables. The key feature is the possibility to create 'what-if' scenarios. The outcomes of these scenarios support the decision-making process by providing an understanding of the problems and the pros and cons of the possible solutions.

8. Acknowledgements

This research has been funded by ProRail and the Next Generation Infrastructure Foundation (NGI). Special thanks go out to the team members both on the side of ProRail (Jelle van Luipen, Emdzad Sehic, the steering committee, amongst many others) and the side of TU Delft (Rens Kortmann, Igor Mayer, Alexander Verbraeck, Bas van Nuland, Gert Jan Stolk and the Game Lab a.o.). Research like this is teamwork.

9. References

Albrecht, T. (2009). The Influence of Anticipating Train Driving on the Dispatching Process in Railway Conflict Situations. In: Networks and Spatial Economics, Springer, vol. 9(1), 2009: pages 85-101.

Barreteau, O. (2003) The joint use of role-playing games and models regarding negotiation processes: characterization of associations. Journal of Artificial Societies and Social Simulation vol. 6, no. 2

Bekebrede, G. (2010). Experiencing Complexity: A Gaming Approach for Understanding Infrastructure Systems. Delft: Next Generation Infrastructures Foundation.

Bekebrede, G. and I.S. Mayer (2005). Build your seaport in a game and learn about complex systems. In: Journal of design research, 5(2): pp 273-298.

Bekebrede, G. and S.A. Meijer (Forthcoming). Gaming for innovation in the health insurance sector.

Botden S.M.B.I., Buzink S.N., Schijven M.P., Jakimowicz J.J. (2008) ProMIS Augmented Reality training of laparoscopic procedures: Face, expert and referent validity. Simulation in Healthcare 3 (2), 97-102

Carter, F.J., M. P. Schijven, R. Aggarwal, T. Grantcharov, N. K. Francis, G. B. Hanna and J. J. Jakimowicz (2005). Consensus guidelines for validation of virtual reality surgical simulators. Surgical Endoscopy. Vol. 19: 12, pp 1523-1532

CBS (2009) Hoe druk is het nu werkelijk op het Nederlandse spoor? Het Nederlandse
    spoorgebruik in vergelijking met de rest van de EU-2. Editors: P. Ramaekers, T. de
    Wit, M. Pouwels. Netherlands Statistics Bureau CBS – in Dutch
De Freitas, S. and M. Oliver (2006). How can exploratory learning with games and
    simulations within the curriculum be most effectively evaluated? Computers and
    Education 46 (3) 249-264
Druckman, D. (1994). The educational effectiveness of interactive games. In D. Crookall and
    K. Arai (Eds) Simulation and gaming across disciplines and cultures. SAGE
    publications, pp. 178-187.
Duke, R.D. (1974). Gaming: the future's language. Sage, Beverly Hills / London.
Duke, R.D. and J.L.A. Geurts (2004). Policy games for strategic management. Dutch
    University Press, Amsterdam, The Netherlands.
Endsley, M.R. (2004). Situation awareness: Progress and directions. In S. Banbury & S.
    Tremblay (Eds.), A cognitive approach to situation awareness: Theory,
    measurement and application (pp. 317–341). Aldershot, UK: Ashgate Publishing.
Gaba, D.M., S.K. Howard, K.J. Fish, B.E. Smith, YA. Sowb (2001). Simulation-Based Training
    in Anesthesia Crisis Resource Management (ACRM): A Decade of Experience.
    Simulation Gaming Vol. 32 no. 2 pp 175-193
Gartner (2011) Case Study: Innovation Squared: The Department for Work and Pensions
    Turns Innovation Into a Game. Accessed through the web 2011-09-30
    http://www.gartner.com/DisplayDocument?id=1476216
Gilbert, N. (2011). Keynote address to Artificial Economics conference, The Hague, The
    Netherlands, 2011-09-01.
Goverde, R.M.P. (2005). Punctuality of Railway Operations and Timetable Stability Analysis.
    TRAIL Research School, Delft, ISBN 90-5584-068-8
Hale, A. and T. Heijer (2006) Is Resilience Really Necessary? The Case of Railway. In:
    Resilience engineering: concepts and precepts. Hollnagel, E., D.D. Woods & Nancy
    Leveson (eds), Ashgate Publishing, Hampshire, UK, pp 125 – 147
Hamilton, W.I. and T. Clarke (2005). Driver performance modelling and its practical
    application to railway safety. Applied Ergonomics Vol. 36: 6, pp 661-670
Hamdorf, J.M. and J.C. Hall (2000): Acquiring surgical skills. British Journal of Surgery. Vol:
    87: 1- pp 28-37
Harteveld, C. (2011) Triadic Game Design: Balancing Reality, Meaning and Play. Springer
Hiltbrand, T. and M. Burke (2011). How Gamification will change Business Intelligence.
    Business Intelligence Journal, Vol 6: 2.
Hollnagel, E., D.D. Woods & Nancy Leveson (2006) Resilience engineering: concepts and
    precepts. Ashgate Publishing, Hampshire, UK.
IEEE (2010) IEEE Recommended Practice for Distributed Simulation Engineering and
    Execution Process (DSEEP), IEEE Std 1730 – 2010, IEEE Explore.
Klabbers, J.H.G. (2003). Simulation and gaming: introduction to the art and science of
    design. Simulation and Gaming, 34(4): 488-494.
Klabbers, J.H.G. (2006). Guest editorial. Artifact assessment vs. theory testing. Simulation &
    Gaming, 37(2): 148-154.
Kortmann, L.J., and E. Sehic (2010) The Railway Bridge Game – usability , usefulness , and
    potential usage for railways management. In proceedings of ISAGA 2010, Spokane.
Kriz, W.C. and J.U. Hense (2006). Theory-oriented evaluation for the design of and research
    in gaming and simulation. Simulation Gaming, 37: 268 – 285.
Mayer, I.S. (2010) The Gaming of Policy and the Politics of Gaming: A Review. Simulation &
    Gaming 40(6) pp. 825–862

Meijer, S.A. (2009) The organisation of transactions : studying supply networks using gaming simulation. Wageningen Academic. ISBN 9789085853343

Meijer, S.A. (forthcoming): Introducing gaming simulation in the Dutch railways. Accepted for TRA 2012, Greece.

Meijer, S.A. and R. Poelman (2011) Supervisor: a 3D serious game for hazard recognition training in the oil industry. Proceedings of ISAGA 2011, Poland.

Meijer, S.A.; Hofstede, G.J.; Beers, G.; Omta, S.W.F. (2006): Trust and Tracing game: learning about transactions and embeddedness in a trade network. Production Planning & Control 17: 6. pp. 569 - 583.

Meijer, S.A., Kracht, P. van der, Luipen, J. van & Schaafsma, A. (2009). Studying a control concept for high-frequency train transport. In E Subrahmanian & J Schueler (Eds.), Developing 21st Century Infrastructure Networks (pp. 1-6). Chennai, India: IEEE.

Meijer, S.A., I.S. Mayer, J. van Luipen & N. Weitenberg (2011): Gaming Rail Cargo Capacity Management: Exploring and Validating Alternative Modes of Organization. Simulation & Gaming 1046878110382161, first published on February 1, 2011 as doi:10.1177/1046878110382161

Middelkoop, A.D. and L. Loeve (2006). Simulation of traffic management with FRISO. Computers in Railways X: Computer System Design and Operation in the Railway and Other Transit Systems.

Noy, A, D.R. Raban and G. Ravid (2006). Testing social theories in computer-mediated communication through gaming and simulation. Simulation & Gaming, 37(2): 174-194.

Peters, V., G. Vissers and G. Heijne (1998). The validity of games. Simulation & Gaming, 29(1): 20 - 30.

Proctor, M.D, M. Bauer, T. Lucario (2007) Helicopter Flight Training Through Serious Aviation Gaming. The Journal of Defense Modeling and Simulation: Applications, Methodology, Technology. Vol. 4:3 pp 277-294

Raser, J.C. (1969). Simulations and society: an exploration of scientific gaming. Allyn & Bacon, Boston.

Rekenkamer (2011) Besteding Spoorbudgetten door ProRail. Report to Parliament by Algemene Rekenkamer, The Netherlands (in Dutch)

Reynolds, A. (1998). Confirmatory program evaluation: a method for strengthening causal inference. The American journal of evaluation, 19: 203 - 221.

Steenhuisen, B., W. Dicke, J.A. de Bruijn (2009). "Soft" Public Values in Jeopardy: Reflecting on the Institutionally Fragmented Situation in Utility Sectors. International Journal of Public Administration, Volume 32, Number 6, May 2009 , pp. 491-507(17)

Van de Velde, D.M., W.W. Veeneman, L. Lutje Schipholt (2008). Competitive tendering in The Netherlands: Central planning vs. functional specifications. Transportation Research Part A: Policy and Practice, Volume 42, 9: pp 1152-1162

Van de Velde, D.M., J. Jacobs, M. Stefanski (2009). Development of railway contracting for the national passenger rail services in the Netherlands. In: International Conference Series on Competition and Ownership in Land Passenger Transport - 2009 - Delft, The Netherlands - Thredbo 11

Van Luipen, J.J.W. and S.A. Meijer (2009). Uploading to the MATRICS: Combining simulation and serious gaming in railway simulators. In: Proceedings of 3rd Int. Conf. on Rail Human Factors, Lille, France, 2009

Witmer B.G., M.J. Singer (1998) Measuring Presence in Virtual Environments: A Presence Questionnaire. Presence Teleoperators Virtual Environments. Vol 7: 3, pp 225-240

## 5

Study and Design of an Electro Technical Device for Safety on Railway Network

Clavel Edith, Meunier Gérard, Bellon Marc and Frugier Didier

G2Elab Grenoble Electrical Engineering Laboratory,
Saint Martin d'Hères, BP46, Saint Martin d'Hères, Cedex,
France

1. Introduction

The security on the railway network is a real important challenge since today, the number of trains is growing and the saturation of the network is close to be reached.

In this chapter, an electrical system is presented in order to improve the electrical detection of trains on the network and correctly manage the lights. That is how security is ensured.

In the next section, the context of the study is presented. The management of lights on network is explained to emphasize the possible trouble which may occur.

The third part deals with the electrical system which is experimented by the French National Railway Company to overcome this problem. The way it works and its main characteristics will be detailed.

In order to base the further developments of this solution not only on experiments, a modeling process is presented in the following part. For other industrial fields, it has proved to be efficient and its use in the case of a railway system seems possible.

In order to validate the modeling approach, an experimental set is developed since it is very difficult to make measurements in situ. It will be presented in the fifth part of this chapter.

Finally in a last section, the results of the modeling process are successfully compared to the measures.

The outlook of improving the studied system is important since the impact of some dimensional parameters on its performances is analyzed whether being geometrical or physical parameters.

2. Context of the study

The distance between trains on a railway is controlled by signals similar to road lights. The railroad line is divided into several sections from 1500m to 20km. Every section is protected by a signal. When a train enters a section ①, an electrical device detects it and makes the light becomes red ②. When the train penetrates into the following section, its signal turns to a red light, while the signal of the first section becomes yellow③. When the train penetrates into the third section, its signal indicates a red light. The signal of the second section gives the indication yellow, and the signal of the first section gives the indication that the way is clear by a green light④. So if a train is stopped in a section, the following train will meet a yellow light announcing to the driver that the following light is red. He will have to reduce the speed of the train to be able to stop if necessary. This is illustrated of Fig. 1.

As said before, an electrical system is used to guarantee this security working. It is constituted by:

- A generator connected to one of the extremities of the isolated section and which imposes a difference of potential between the two rails,
- A receptor connected to the other extremity of the isolated section which measures the difference of potential between the two rails,
- A transmission line which is electrically limited to the extremities of the considered section.

The generator sends a coded signal through the electric circuit constituted by the rails and the receptor. When the zone included between the generator and the receptor is free, the receptor is able to detect the coded signal: the way is then considered free. When a train enters this section, a great part of current is derived by wheels and axles (what we call "shuntage"); the receptor does not receive this coded signal coming from the generator anymore: the way is then considered busy. Such a device is thus able to detect the presence of trains on a section by the change of the impedance value of the circuit between the rails. When a train is on a section, the rails are short-circuited by the train and the group wheels, axles, rails 1 and 2 have lower impedance. So the measure of the voltage drop at the receptor implies a busy section. This change of electrical circuit is illustrated on Fig. 2.

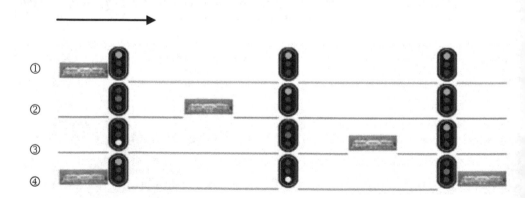

Fig. 1. The principle of railway lights

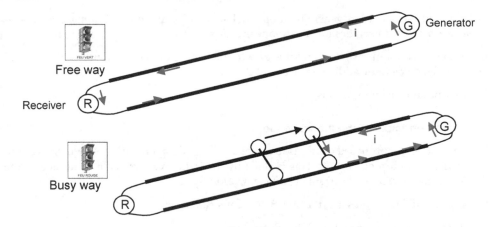

Fig. 2. The principle of the detecting electrical system

In the case of the electrified lines, which represent 90 % of the French network, rails are very often used to ensure the return of drive current towards the substations. Drive currents are about 1000A. And they have to coexist with currents of 1A from the detecting electrical system.

In order to be able to easily separate these two currents, the kind of current inside the detecting system depends on the electrified lines. In the case of DC lines, alternative currents are used for the detecting electrical system and in the case of alternative power supply different frequencies are used for the detecting circuit. In that case, the insulation between two detecting systems is carried out using inductive connections which allow the return of the drive current.

Current of the detecting circuits can be modulated with various frequencies, which can be detected by equipments on board the trains to give to the drivers all the indications in the cabin. This principle of the transmission between the way and the machine is used in the famous French High Speed Train (TGV).

Sometimes, the short circuit between wheels and rails is not of good quality. This is the case of weak machines, parts of the railway network rarely used, bad wheel/rail contact due to insulating body. So errors of detection by the system can lead to dangerous situations. The light is green even if there is a train in the section. If a more rapid train arrives, an accident could occur.

As a consequence, in order to be sure that the signal is well interrupted by the train on a section, the French National Railway Company (SNCF) has added an electrical device which will be described in the following part.

3. The proposed device

To avoid this problem of detection, several solutions have been explored:

- to replace all the detecting electrical systems,
- to improve the contact between the wheels and the rails.

This last solution has been generally adopted. This can be made by cleaning the rails. A special accessory has been designed to scrape the rails.

But another way to improve the contact is to help the short circuit created by the wheels and the rails to be efficient by adding an electrical device.

This last option is the topic of this paper.

3.1 Working principle of the inductive loop

The principle of the proposed device is to create a voltage drop of sufficient level in order to make possible a current to flow. This is done by an inductive loop able to induce in the wheels/axles/rails circuit a current of sufficient value.

The proposed device is presented on Fig. 3. It is constituted by:

- a high frequency current controlled generator,
- a parallel LC circuit made of capacitors located on the bogies and the inductance of the inductive loop.

This loop is in fact the primary circuit of a transformer. The secondary part is constituted by the axles and the rails. The current induced by the loop flows through the axles and the rails and must be high enough to guarantee the electrical contact between the wheels and the rails. During a bad contact rail / wheel, the secondary circuit of the transformer is open and an induced voltage appears between the wheel and the rail and establishes again the electrical contact insuring the "shuntage".

The chariot near the inductive loop constitutes a third circuit of the transformer which can reduce the value of the current inside the axles – rails circuit.

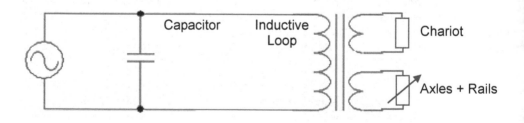

Fig. 3. The principle of the detecting electrical system

3.2 Electrical and mechanical requirements

To be efficient, this system has to create a minimum voltage drop between a wheel and the rail. The working frequency of this system is imposed by the French National Railway Company. The design data for the loop are:

- minimum voltage drop : 3V,
- frequency : 147kHz,
- range of frequencies : 140kHz to 152kHz (adjusted using capacitors),
- control of the current : if the current remains under a certain limit during 10s, a default signal is sent to the driver. An emergency braking is used and special rules are adopted by the driver,
- mechanical constraints (under the train, easy to access, easy to change and repair,…),
- material for the loop : copper is used and several other cables are added.

3.3 Design of the loop

The performances of the loop are directly linked to its geometrical size.

But the degrees of freedom are limited because a lot of mechanical constraints exist:

- the surface of the loop is limited by the space under the train,
- the height of the loop is limited by the distance between the train and the rails,
- leakage currents are flowing inside the metallic structure.

In practice, the environment of the loop changes (shunt, ballast, metallic parts …) and the value of the inductance of the loop changes too. So the resonance frequency is modified. In order to avoid this, it is necessary to continuously adjust the frequency so that the current is sufficient inside the loop.

In order to perform this, a Phase-locked loop (PLL) is used.

Thanks to three electronic cards, every change of frequency is detected and automatically compensated:

- a ALIM card allows the galvanic insulation and adapts the voltage which supplies the system (24V),
- a BIAS card controls the value of 147kHz,
- a TFER card is a controlled amplifier for the BIAS card.

In order to reach an optimal design of the loop, experiments are not sufficient.

A modeling process has to be run in order to take into account all the requirements and desired performances.

This is the aim of the following part: to present a modeling approach.

4. Modeling method

As said before, the aim of the modeling process is to be able to evaluate the voltage and current in the studied structure. This implies to solve the Maxwell's equations and to establish an electrical equivalent circuit on which the circuit equations could be solved. In the field of electromagnetic approach, two families of modeling methods are facing:

- the finite element method (FEM),
- the integral method.

4.1 Choice of the modeling method

Each family of modeling method has advantages and drawbacks.

Concerning the FEM, it is well known and gives good results on the evaluation of the electromagnetic fields everywhere in the space. But it requires the meshing of all the space, i.e. the studied structure but also the air around. Moreover good results imply a good use of the formulations and assumptions at the limits of the studied domain. The results are principally the electromagnetic fields. Then to obtain the electrical equivalent circuit, post-processing evaluations have to be done.

Concerning the integral methods, the only meshing of the conductive parts makes them very attractive. The number of unknowns is limited. Moreover, an electrical equivalent circuit can directly be deduced from the solving.

The most famous integral methods are the method of moment (MoM) and the PEEC method (Partial Element Equivalent Circuit).

Since the studied structure could be very large and the amount of air around significant, the use of FEM could lead to solve problems with too high a number of unknowns.

So an integral method is chosen to model the studied structure and more particularly the PEEC method which is detailed in the following paragraph.

4.2 Principle of the PEEC method

The PEEC method was firstly introduced by A. Ruehli (Ruehli, 1974). Based on low frequency exact analytical formulae, it consists in extracting the electric parameters from the geometries of conductors. This allows taking into account the electrical parasitic effects of interconnections while evaluating the electromagnetic behavior of an electronic or electrical system.

Full PEEC method takes into account resistive, inductive and capacitive parts.

4.2.1 Assumptions

According the frequency range of the study, the electrical equivalent model could be more or less complicated. Indeed for not so high frequencies, only resistive and inductive effects of cabling are involved in the electromagnetic behavior of the studied system. That is why it is possible to partly use the PEEC method which allows reducing the size of the equivalent model. For the present studied application, frequency is low enough (around some kHz) to limit the evaluation to the resistive and inductive parts.

Nevertheless the capacitive aspect is detailed in (Ardon et al., 2009) to complete the study.

The mains equations will be detailed in the following paragraph to obtain the equation (13) which is the base to establish the electrical equivalent circuit.

In order to easily solve the equation (13) for each considered part, the most important assumption is that current density is uniform. But according the frequency and geometry of the studied structure, skin and proximity effects have to be taken into account during the modeling. So in a first step, all the conductors of the studied structure have to be meshed so that the current density is uniform inside each mesh.

Moreover, as presented in (13), relative permeability μ_r is assumed to be equal 1. Otherwise further developments of PEEC method are presented and detailed in (Aimé et al., 2009b) and (Kéradec et al., 2005) in order to take into account magnetic material influence on current distribution.

Finally no propagation aspect is considered for this first presentation. Otherwise, another modeling method has to be applied such as rPEEC or transmission lines approach (Antonini et al., 2007) and (Clavel et al., 2007).

4.2.2 Equations and associated model

The problem can be better formalized by considering the integral form of the Maxwell's equations and by assuming:

- quasi-static conditions;
- only surface location for the free-charges ρ;
- uniform and constant currents I in each volume element of conductors;
- constant material conductivity σ,
- vacuum permeability μ_0 surrounding the objects;
- a homogeneous medium of permittivity $\varepsilon = \varepsilon_r \, \varepsilon_0$;
- negligible losses in dielectric materials.

In such conditions, the following equations can be written:

$$\mathbf{rotH} = \mathbf{J} \tag{1}$$

$$\mathbf{rotE} = -j\omega\mathbf{B} \tag{2}$$

$$div\mathbf{B} = 0 \tag{3}$$

$$\mathbf{B} = \mu_0.\mathbf{H} \tag{4}$$

$$\mathbf{J} = \sigma\mathbf{E} \tag{5}$$

From (3), it comes:

$$\mathbf{B} = \mathbf{rotA} \tag{6}$$

$$\mathbf{E} = -j\omega\mathbf{A} - \mathbf{grad}V \tag{7}$$

$$\Delta\mathbf{A} = -\mu_0\mathbf{J} \tag{8}$$

(7) is the Faraday's law and (8) the Poisson's equation.

Considering a conducting volume Ωc in an external electrical field (V_{ext}), the total electrical field E_{total} at any point P in the conductor is given by (9):

$$\mathbf{E}_{total}(P) = -j\omega\mathbf{A}(P) - \mathbf{grad}\, V_{charge}(P) - \mathbf{grad}\, V_{ext}(P) \tag{9}$$

V_{charge} is the potential due to the electrical charges in the conductor.

If, in a first approximation, the capacitive effects can be neglected, this term is null.

From (8), without propagation effects and magnetic material, it comes (10):

$$\mathbf{A}(P) = \frac{\mu_0}{4\pi} \int_{\Omega c} \frac{\mathbf{J}}{r} d\Omega \tag{10}$$

where J is current density and r the distance between the integration point and P.

Taking (5) and (10) in (9) gives (11):

$$\frac{\mathbf{J}(P)}{\sigma} + j\omega\frac{\mu_0}{4\pi} \int_{\Omega c} \frac{\mathbf{J}}{r} d\Omega = -\mathbf{grad}\, V_{ext} \tag{11}$$

In order to reach the desired model, it is necessary to suppose a uniform current density. For that, the volume Ωc is divided into m elementary conductors. On each elementary conductor, (12) is written:

$$\mathbf{J}_k = \mathbf{J}_{0k}.I_k \tag{12}$$

J_{0k} is the electrokinetic solution for a 1A current.

Multiplying (11) by J_{0k}, it comes (13):

$$\int_{\Omega c_k} \mathbf{J}_{0k} \cdot \frac{\mathbf{J}_k}{\sigma} d\Omega c_k + j\omega\frac{\mu_0}{4\pi} \int_{\Omega c_k} \mathbf{J}_{0k}\left(\sum_{i=1}^{m} \int_{\Omega c_i} \frac{\mathbf{J}_{0i}.I_i}{r} d\Omega c_i \right) d\Omega c_k = -\int_{\Omega c_k} \mathbf{J}_{0k}\mathbf{grad}\, V_{ext}.d\Omega c_k \tag{13}$$

From (13), the electrical equivalent circuit of a conductor can be deduced (Fig. 4) (Ruehli & Cangellaris, 2001).

(12) inside the first term of (13) gives (14):

$$\int_{\Omega c_k} \mathbf{J}_{0k} \cdot \frac{\mathbf{J}_k}{\sigma} d\Omega c_k = \frac{1}{\sigma} I_k \int_{\Omega c_k} \mathbf{J}_{0k}^2.d\Omega c_k \tag{14}$$

Knowing (15), (16) can be deduced.

$$\frac{1}{\sigma} \int_{\Omega c_k} \mathbf{J}_{0k}^2.d\Omega c_k = R_k \tag{15}$$

$$\int_{\Omega c_k} \mathbf{J}_{0k} \cdot \frac{\mathbf{J}_k}{\sigma} d\Omega c_k = R_k.I_k \tag{16}$$

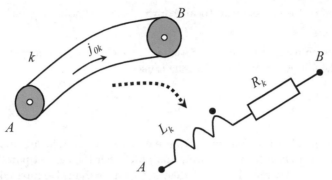

Fig. 4. k^{th} element of the conductor

Introducing the mutual inductance between two elementary conductors k and i (17), the second term of (13) leads to (18).

$$L_{ki} = \frac{\mu_0}{4\pi} \int\limits_{\Omega c_k} \int\limits_{\Omega c_i} \frac{\mathbf{J}_{0k}.\mathbf{J}_{0i}}{r} d\Omega c_i d\Omega c_k \tag{17}$$

$$j\omega \frac{\mu_0}{4\pi} \int\limits_{\Omega c_k} \mathbf{J}_{0k} \left(\sum_{i=1}^m \int\limits_{\Omega c_i} \frac{\mathbf{J}_{0l}.I_i}{r} d\Omega c_i \right) d\Omega c_k = j\omega \sum_{l=1}^m L_{kl}.I_l \tag{18}$$

Concerning the term on the right of (13) the following equations can be established assuming the electrokinetic solution gives $div\mathbf{J}_0 = 0$:

$$div(V.\mathbf{J}_{0k}) = \mathbf{J}_{0k}.\mathbf{grad}\,V + V.div\mathbf{J}_0 \tag{19}$$

$$\int\limits_{\Omega c_k} \mathbf{J}_{0k}\mathbf{grad}\,V_{ext}.d\Omega_k = \int\limits_{\Omega c_k} div(V_{ext}.\mathbf{J}_{0k})d\Omega_k = \oint\limits_{\Gamma_k} V_{ext}.\mathbf{J}_{0k}.\mathbf{n}_k.d\Gamma_k \tag{20}$$

$$\mathbf{J}_{0k}.\mathbf{n}_k = 0 \tag{21}$$

$$\oint\limits_{\Gamma_k} V_{ext}.\mathbf{J}_{0k}.\mathbf{n}_k.d\Gamma_k = V_a \int\limits_{\Gamma_a} \mathbf{J}_{0k}.\mathbf{n}_a.d\Gamma_k + V_b \int\limits_{\Gamma_b} \mathbf{J}_{0k}.\mathbf{n}_b.d\Gamma_k = -V_a + V_b = -U_k \tag{22}$$

with Γ_k is the edge of volume Ωc_k; $\mathbf{n}_k$ is the normal vector oriented towards the exterior of the surface, Γ_a and Γ_b are the current outputs surfaces.

Hence, the electrical equation is deduced (23) and (24):

$$- \int\limits_{\Omega c_k} \mathbf{J}_{0k}\mathbf{grad}\,V_{ext}.d\Omega c_k = U_k \tag{23}$$

$$R_k.I_k + j\omega \sum_{i=1}^m L_{ki}.I_i = U_k \tag{24}$$

So each term of (13) leads to an electrical characteristic of conductor which is only function of its geometry.

To compute the parasitic resistance R_i in each volume element V_i of length ℓ_i, section S_i, and resistivity ρ, the following analytical formula is used:

$$R_i = \rho \frac{l_i}{S_i} \qquad (25)$$

Because of the parallelepiped shape of the elements the double integral in (17) can be expressed in an analytical form and easily computed. For the case of parallel elements, the mutual inductances M_{ij} are computed thanks to an analytical expression. And for the general case presented on Fig. 5, the expressions (26) can be used either to evaluate M_{ij} but also partial inductance L_i if a=d, b=c, l1=L2 and E=p=l3=0 (Hoer & Love, 1965). If elements are not parallel an analytical/numerical integration technique is used (an analytical expression for the first integral is used, the second one being computed thanks to an adaptive gauss point integration ensuring a good accuracy). All values of L_i and M_{ij} can then be organized in a dense and square matrix [L-M] whose size is equal to the number of mesh elements (Aimé et al., 2007).

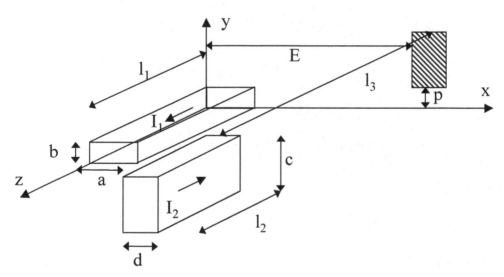

Fig. 5. General case of parallel elements for the evaluations of M_{ij}

$$M_{ij} = \left[\left[\left[f(x,y,z)\right] \begin{matrix} E+d,E-a \\ (x) \\ E+d-a,E \end{matrix} \right] \begin{matrix} p+c,p-b \\ (y) \\ p+c-b,p \end{matrix} \right] \begin{matrix} l3+l2,l3-l1 \\ (z) \\ l3+l2-l1,l3 \end{matrix} \qquad (26)$$

$$\left[\left[\left[f(x,y,z)\right] \begin{matrix} x_1,x_3 \\ (x) \\ x_2,x_4 \end{matrix} \right] \begin{matrix} y_1,y_3 \\ (y) \\ y_2,y_4 \end{matrix} \right] \begin{matrix} z_1,z_3 \\ (z) \\ z_2,z_4 \end{matrix} = \sum_{i=1}^{4}\sum_{j=1}^{4}\sum_{k=1}^{4} (-1)^{i+j+k+1} f(x_i,y_j,z_k)$$

$$f(x,y,z) = \frac{\mu_0}{4\pi} \frac{1}{a.b.c.d} \left(\begin{array}{l} \left(\dfrac{y^2 z^2}{4} - \dfrac{y^4}{24} - \dfrac{z^4}{24} \right) x \ln\left(x + \sqrt{x^2 + y^2 + z^2} \right) + \\[2mm] \left(\dfrac{x^2 z^2}{4} - \dfrac{x^4}{24} - \dfrac{z^4}{24} \right) y \ln\left(y + \sqrt{x^2 + y^2 + z^2} \right) + \\[2mm] \left(\dfrac{y^2 x^2}{4} - \dfrac{y^4}{24} - \dfrac{x^4}{24} \right) z \ln\left(z + \sqrt{x^2 + y^2 + z^2} \right) + \\[2mm] \dfrac{1}{60}\left(x^4 + y^4 + z^4 - 3y^2 x^2 - 3y^2 z^2 - 3x^2 z^2 \right)\sqrt{x^2 + y^2 + z^2} - \\[2mm] \dfrac{xyz^3}{6}\text{Arctan}\dfrac{xy}{z\sqrt{x^2 + y^2 + z^2}} - \dfrac{xy^3 z}{6}\text{Arctan}\dfrac{xz}{y\sqrt{x^2 + y^2 + z^2}} - \\[2mm] \dfrac{x^3 yz}{6}\text{Arctan}\dfrac{zy}{x\sqrt{x^2 + y^2 + z^2}} \end{array} \right)$$

The evaluation of the electrical equivalent circuit of the meshed structure has been implemented into the dedicated software InCa3D® which offers a robust and fast solver combined to a user friendly and efficient graphical interface.

For simple shapes of elements, analytical formulations (25) and (26) are used. But if the geometrical configuration is more complex, a numerical technique is used to find the values of the equivalent circuit.

4.2.3 Meshing techniques

In order to properly describe the current distribution inside conductors, an adapted meshing technique has to be applied. Indeed the shape of conductors often allows the assumption of 1D or 2D current flowing inside them.

For massive bars, or cables the 1D current leads to only mesh the cross section of conductor. Since no propagation effect is described, the length of conductors has not to be subdivided (Fig. 6). For this kind of conductor, the skin effect can be taken into account concentrating the meshes on the edges of the conductors so that the number of subdivisions is not too big.

PEEC
elements

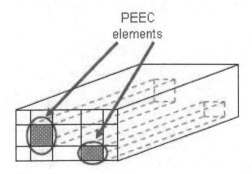

Fig. 6. The 1D meshing of conductors

In the case of very thin and large conductors such as sheet of copper, ground plane, the 1D current assumption is no more valid. Indeed current is generally flowing in a plane so that a 2D approximation can be sufficient in order to properly describe the physical phenomenon. Two quadrate directions for current inside the conductor are defined. So the developed 2D meshing technique consists in dividing the plane as presented on Fig. 7.

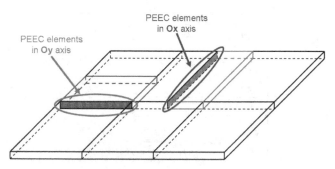

Fig. 7. The 2D meshing of conductors

For both cases, according to the shape of the cross section or conductor, a refinement meshing technique is applied so that the description is close to the real structure (Fig. 8).

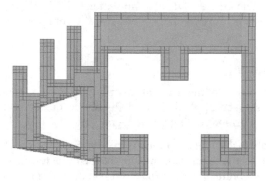

Fig. 8. Examples of meshing for complex cross section for 1D assumption

The associated electrical equivalent models for the 1D and 2D elements are summarized on Fig. 9.

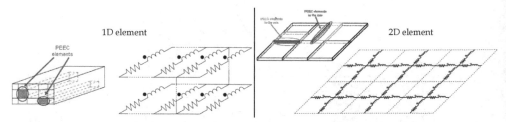

Fig. 9. R-L-M electrical equivalent circuit for 1D and 2D elements (mutual inductances are not represented on the figure)

4.3 Modeling process

On Fig. 10, the modeling process of every kind of electrical structure is detailed.

It consists of four steps:

- Geometry description from designers' data or directly imported from CAD tools,
- Meshing according the appropriated assumption for the current,
- Solving the equations to obtain the electrical equivalent circuit,
- Circuit solving.

For the studied case, the unknowns are the currents inside the equivalent circuit. For that purpose it is necessary to describe the electrical environment of the problem in order to solve the right circuit equations. This last step can be achieved in a circuit solver like SPICE® or Portunus® exporting automatically the equivalent circuit inside these tools. But the size of the equivalent circuit is linked to the number of meshes and can be too big for these tools. Moreover it is not necessary to keep the information of local current inside each mesh. What is interesting is the global current inside the conductors. So a reduced equivalent circuit is better appropriated for this goal. To that aim, the user has to clearly identify the outputs of each conductor and then using parallel and series associations, the equivalent impedance between these points can be evaluated for each frequency. This reduced circuit is afterwards more practical in order to evaluate all necessary currents and voltages. Even if it is frequency dependent, a time simulation can be efficiently done. Indeed, according the frequency range, this dependence can be negligible and if not, numerical techniques to find a non dependent circuit with more components exist (Tan & He, 2007).

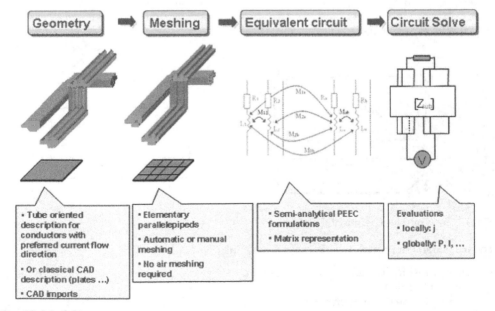

Fig. 10. Modeling process

This proposed modeling process has proved to be very efficient not only for high current electrical systems (Gonnet et al., 2004) but also power electronics devices and structures (Aimé et al., 2009a), electronic card (Clavel et al., 2007a), PCB application (Vialardi et al., 2010) and aircraft structures (Jazzar et al., 2011).

5. The experimental structure

In order to validate the modeling process, experiments have been undertaken.

Unfortunately, measures on real structures are very tricky. A simplified system has thus to be defined. It is presented on Fig. 11. The chariot has been replaced by copper tubes (diameter 42mm); the rails and axles have been replaced by cables (35 mm^2). Sizes have been chosen so that the experimental set is close to a real system (Fig. 12).

The characteristics of the voltage source are the same as described in the requirements paragraph.

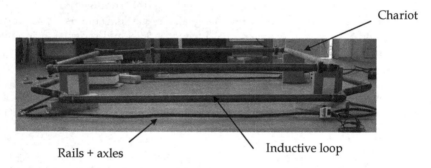

Fig. 11. The experimental system

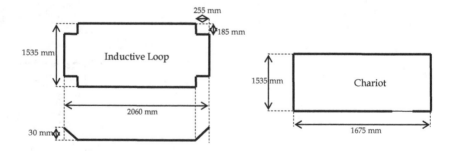

Fig. 12. Sizes of the experimental system

The measurement procedure consists in measuring (Fig. 13-14):

- current inside the chariot (so the cross section has been locally reduced in order to insert a current probe);
- current inside the inductive loop;
- voltage drop between the wheel and the rails (Urw).

Different configurations have been tested and for each of them maximum data have been measured:

- the inductive loop alone;
- the inductive loop + rails + axles;
- the inductive loop + rails + axles + chariots;
- the inductive loop + rails + axles + chariots + 4 shunts for the wheels;
- the inductive loop + rails + axles + chariots + 4 shunts for the wheels with 4 resistors to represent the non perfect electrical contact between the wheel and the rail.

For each of them open circuit and short circuit measurements have been made.

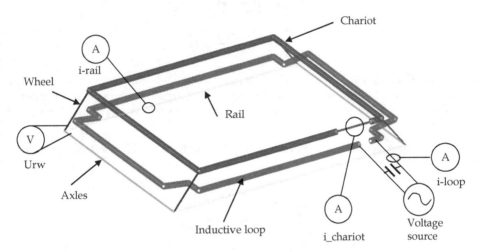

Fig. 13. Measurements procedure

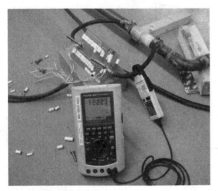

Current measure

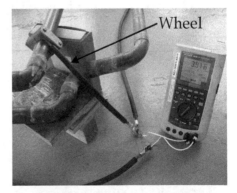

Voltage drop measure

Fig. 14. Used probes (Fluke 867B, current probe PR30)

Knowing the current I in the loop, the voltage source V, their phase φ and the frequency (f, ω=2πf) an equivalent circuit for the measured system can be calculated using (27) (Fig. 15).

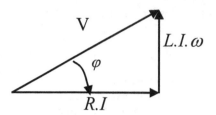

Fig. 15. Fresnel's diagram for the evaluation of equivalent circuit

$$\begin{cases} R = \dfrac{V \cos(\varphi)}{I} \\ L = \dfrac{V.\sin(\varphi)}{2\pi f.I} \end{cases} \tag{27}$$

The second configuration is very close to the classical case of two coupled inductances like a simple transformer. The first one is supplied with an alternative source and an induced current (short-circuit situation) or voltage (open circuit situation) is created on the second one (Fig. 16).

The theoretical study is briefly reminded in the following equations (28) with only one turn for our case (n1=n2=1).

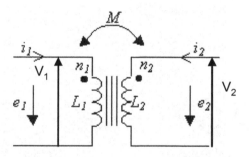

Fig. 16. The simple transformer

$$V_1 = R_1.I_1 + L_1 \frac{dI_1}{dt} + M \frac{dI_2}{dt}$$
$$V_2 = R_2.I_2 + L_2 \frac{dI_2}{dt} + M \frac{dI_1}{dt} \tag{28}$$

The open circuit configuration gives (29). The voltage drop is directly the Urw voltage characterizing the working of the loop.

$$V_{10} = R_1.I_{10} + L_1 \frac{dI_{10}}{dt} = R_1.I_{10} + L_1.I_{10}.\omega$$
$$V_{20} = M \frac{dI_{10}}{dt} = M.I_{10}.\omega \tag{29}$$

6. Results

In this part the electrical equivalent circuit given by the modeling process is presented as well as comparisons between the circuit simulation and measurements.

6.1 Modeling of the real system

The CAD complete studied structure is presented on Fig. 17 on which the wheels, chariot, inductive loop clearly appear.

Indeed in order to correctly solve the problem, some geometrical simplifications have been done. The wheels, the rails and the axles have been replaced by straight massive conductors; four shunts have been added to connect the chariot to the axles in order to represent a realistic situation.

Using InCa3D, the result is presented on Fig. 18.

The geometry has been meshed in order to take into account the proximity and frequency effects.

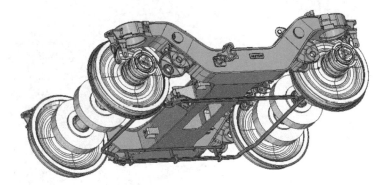

Fig. 17. The complete structure

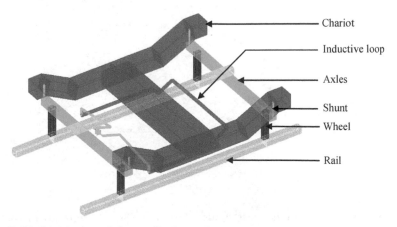

Fig. 18. InCa3D description of the studied structure

After the PEEC solving, the equivalent circuit has been reduced in order to obtain the simple SPICE-like circuit drawn on Fig. 19 where each part of the system is well identified by an L-R series equivalent circuit. On this circuit all the inductances are coupled with mutual coefficients but they have been cut off to make it clearer.

6.2 Comparison between measures and simulation on the experimental set

The experimental case has been modeled (Fig. 11) using the same process and for each configuration simulation results with the same operating conditions (value of the supply voltage source, frequency) have been compared to the measurements.

The results are presented in Table 1 for the closed circuit and Table 2 for an open one.

A good agreement between simulations and measurements can be observed.

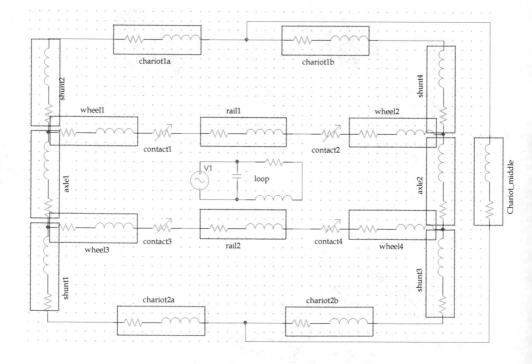

Fig. 19. Reduced electrical equivalent circuit for the studied structure (mutual inductances have been removed)

Conditions		Closed circuit			
		I_loop (A)	Φ (°)	L (μH)	R (Ω)
Only the loop	Measure	10.07	-98	5.15	0.25
	Simulation	10.72	-89.9	4.84	0.01
Loop + Rails + Axles	Measure	11.13	-96	4.5	0.44
	Simulation	12.84	-89.8	3.98	0.01
Loop + Rails + Axles + Chariot	Measure	12.6	-95.8	4.05	0.38
	Simulation	14.65	-89.8	3.5	0.01
Loop + Rails + Axles + Chariot + 4 shunts	Measure	12.5	-95	4.16	0.34
	Simulation	14.95	-89.8	3.49	0.01

Table 1. Current in the loop and electrical characteristics for a closed circuit - simulation and measurements

Conditions		Open circuit					
		I_loop (A)	Φ (°)	L (μH)	R (Ω)	Urw (V)	Δurw
Loop + Rails + Axles	Measure	10.8	-95.9	5.21	0.46	24.1	4.66 %
	Simulation	11.68	-89.8	4.84	0.01	25.22	
Loop + Rails + Axles + Chariot	Measure	12.04	-95.5	4.59	0.38	21.67	5.45 %
	Simulation	13.8	-89.8	4.02	0.01	22.85	
Loop + Rails + Axles + Chariot + 4 shunts	Measure	12.3	-95.3	4.35	0.37	8.4	1.9%
	Simulation	14.7	-89.8	3.65	0.01	8.56	

Table 2. Rail-Wheel voltage and electrical characteristics for an open circuit - simulation and measurements

6.3 Parametric analysis

Once the modeling process is established with satisfactory results, it is possible to make some changes to analyze the influence of some parameters on the performances of the system.

Indeed, using InCa3D, it is possible to define geometrical and physical parameters and make them varying to improve the design of the inductive loop.

For the studied structure, the following characteristics can be defined as parameters:

- height of the inductive loop;
- height of the chariot;
- size of the loop;
- position of the loop under the train;
- number of shunts;
- diameter of the loop;
- number of turns of the loop;
- resistivity of the material;
- voltage source;
- frequency;
- resistance of the wheel/rail contact.

The main performance is the rail/wheel voltage Urw.

6.3.1 Variation of the height

The distance between the inductive loop and the rails is varying in this study.

On Fig. 20, results show that the lower this distance is, the higher the Urw is.

This result is quite logical since according to (30) the induced voltage is linked to the magnetic flux. And when the distance between the two loops is low this flux is maximal.

$$e = -\frac{d\varphi}{dt} \tag{30}$$

Nevertheless, the range of variation of this parameter is quite limited because mechanical constraints on the train and required standards for some sizes. There is a minimum value to respect.

By way of contrast, if the distance between the loop and the chariot decreases, the value of Urw decreases too. This is due to the fact that, as said before, the chariot creates a supplementary winding (Fig. 3) in which an induced current can be created. The inductive coupling in this case is higher than between the loop and the rails.

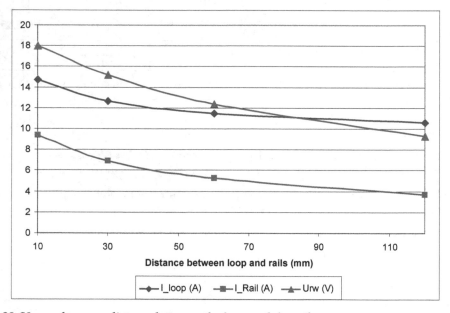

Fig. 20. Urw voltage vs. distance between the loop and the rails

6.3.2 Variation of the sizes and the position of the loop

On Fig. 21, three configurations have been modeled in order to evaluate the impact of the sizes and the position of the loop.

Along the simulations, it has been observed that it is essential that the loop is parallel to the rails and axles unless it performance drastically decreases.

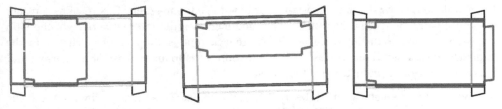

Fig. 21. Three configurations for the sizes and the position of the loop

For example, by reducing by half the length of the loop, the Urw decreases by 32.8%. And by reducing by half the width of the loop, the Urw decreases between 26% and 52% (depending on the place where it is measured -close to the loop or not). If, on the contrary, the length of the loop is increased of 25%, Urw decreases by 11.6%.

The results obtained during these simulations confirm that the coupling between the primary constituted by the loop, and the secondary constituted by the circuit rails-axles, depends on the geometry of the loop. The purpose is obviously that the secondary circuit shares as many lines of magnetic field as possible with the primary one. Since there is no magnetic material which can guide the magnetic flux, contrary to a classic transformer, it is necessary to move as close as possible from the inductive loop of the secondary conductors to increase the coupling.

Keeping the same loop and moving it under the train so that it is no more under the chariot, the distance between the primary part and the secondary one will increase and the performances will decrease. A secondary part constituted by both rails and axles of two different chariots can also be imagined. Yet the distance between the axles of two consecutive chariots is approximately 15 meters. The length of the loop is then considerably increased, and thus its inductance (ratio 3.7). The voltage source of the loop would then be oversized to obtain an Urw voltage equivalent to that of the current loop. Moreover the cost of the loop would be drastically increased and additional mechanical constraints to fix the loop would appear.

6.3.3 Variation of the diameter of the loop

Keeping the same external sizes for the loop, the diameter of the copper tube has been increased in this part of the study. As presented on Fig. 22, this implies a reduction of the internal surface of the loop.

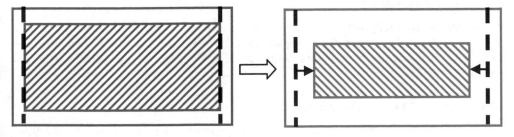

Fig. 22. Increase of the diameter f the loop

The simulations show that for a constant supply voltage the magnetic flux remains constant too. Since $\Phi=BS$, when the diameter of the tube increases the internal surface of the loop decreases. So the induction B increases. As the current is directly proportional to the induction and as $\Phi=LI$, the increase of the induction leads to an increase of the current and thus a decrease of the inductance of the system.

By dividing the diameter of the tube by 2 (21mm), the surface increases by 7.5% and the current in the loop decreases by 13.6%. The inductance increases by 16.7%. The result of this variation of diameter is the decrease of Urw voltage by 8%.

On the contrary, if the diameter of the tube is doubled (84mm), the surface decreases by 13.8%, the current in the loop increases by 20%. The Urw voltage increases by 8%.

But it is not possible to increase this parameter too much because the increase of the current would imply more Joule losses and a higher temperature.

At this step, a question could be to determine the maximum value for the current in the loop regarding the losses but also the EMC performances of the whole system.

6.3.4 Variation of the number of turns for the loop

Keeping the same external shape as the current loop, a second turn has been added using a copper tube of 20mm of diameter.

This additional turn considerably increases the inductance (ratio 3.7). Theoretically, the inductance is proportional to the square of the number of turn (4 in this case).

So with a constant voltage source, the current inside the loop thus decreases with the same ratio and the created magnetic field is also weaker. The Urw voltage is then decreasing by 46%. So it is necessary to increase the voltage source to keep the same performances.

6.3.5 Variation of the number of the shunts

On Fig. 19, it is clear that the four shunts, which connect the chariot to the axles to ensure a return path for the current and protect the persons of an electric risk, introduce new loops in which current can flow.

Using simulations it is possible to connect -or not- these shunts in order to evaluate their impact on the Urw voltage.

Indeed, considering only two shunts Urw voltage increases by 110%. And with only one shunt, no induced current can flow since the circuit is open and the Urw voltage increases then by 210%.But since these shunts ensure the electric safety of the persons, it is not possible to eliminate them.

6.3.6 Variation of the resistivity of the material

Considering the results presented in Table 1, the impedance of the loop is mainly inductive , all the more because the resistive part becomes negligible since the frequency is high.

So by changing the resistivity of material, aluminum instead of copper (ratio 1.6), no impact on the Urw voltage can be underlined.

This aspect is very interesting in an economic point of view because the loop could be made of a material cheaper than copper.

6.3.7 Variation of the electrical contact

To model the electrical contact between the wheel and the rail a resistor has been added because this is not a perfect short-circuit (Fig. 19).

By making the value of this resistor vary, it is possible to deteriorate the contact.

Results are presented on Table 3 for a resistance varying from 1Ω to $1M\Omega$.

For values higher than 5Ω, the Urw voltage remains constant.

And for a 42V voltage source, the Urw is about 3.7V. This characteristic has been drawn on Fig. 23 and is close to that of the voltage at the secondary winding of a transformer.

This study shows that even if the four wheels of the chariot are not in electric contact with the rails, as with a resistance of contact of $1M\Omega$, the inductive loop always allows to obtain a sufficient Urw voltage which corresponds in this case to the open circuit voltage of a transformer.

The bad quality of contacts which is one of the causes of non detection of trains inside a section is then swept away.

R	I loop	I Rail	Urw
$0\,\Omega$	12.9 A	2.55 A	4.2 V
$1\,\Omega$	12.57 V	1.50 A	3.96 V
$5\,\Omega$	12.42 V	0.367 A	3.73 V
$10\,\Omega$	12.42 A	0.185 A	3.68 V
$100\,\Omega$	12.41 V	$1.85\ 10^{-2}$ A	3.68 V
$1\,k\Omega$	12.41 V	$1.85\ 10^{-3}$ A	3.68 V
$10\,k\Omega$	12.41 V	$1.85\ 10^{-4}$ A	3.68 V
$50\,k\Omega$	12.41 V	$3.71\ 10^{-5}$ A	3.68 V
$100\,k\Omega$	12.41 V	$1.85\ 10^{-5}$ A	3.68 V
$500\,k\Omega$	12.41 V	$3.71\ 10^{-6}$ A	3.68 V
$1\,M\Omega$	12.41 V	$1.85\ 10^{-6}$ A	3.68 V

Table 3. Comparison between simulation and measurements for an open circuit

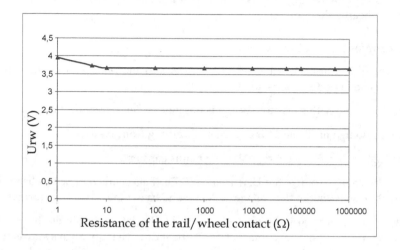

Fig. 23. Urw vs. resistance of the rail/wheel contact

6.3.8 Variation of the frequency of the source

The supply source of the inductive loop is a 147kHz sinusoidal voltage which is obtained from the 72V battery embedded on the train and electronic cards. This is a specific source and no particular constraints are linked to the other electric systems embedded on the train. So it is possible to imagine a variation of the level of voltage as well as the frequency.

It is clear that, the Urw voltage is directly linked to the voltage value (proportionality). This is logical since no magnetic material has been taken into account in this study.

If there is a magnetic material, with the increase of the voltage, saturation will appear and this proportionality relation will be wrong.

Concerning the frequency, using simulations, it has been changed from 50Hz to 1MHz.

On Fig. 24 the frequency evolution of the Urw voltage is represented.

With the obtained results, since the resistive part of the loop is very low, the famous relation $U = L\omega I$ even for low frequencies is valid. So the current into the loop could be very high for low frequency and but could decrease with increasing frequency.

And to ensure a sufficient value for the Urw voltage it must be higher than some 10A.

So compromise has to be reached to improve the inductive loop, between the values of the voltage source, its frequency, losses, possible saturation if magnetic materials are used.

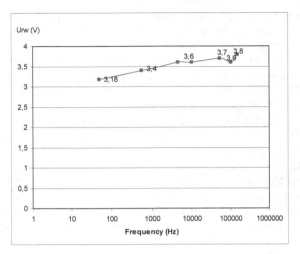

Fig. 24. Urw (V) vs. frequency (Hz) of the voltage source

7. Conclusions

The number of trains on the national network is increasing. In order to ensure a maximum security, it is necessary to localize all the trains all the time.

For that purpose, the French National Railway Company (SNCF) uses an electronic detection based on the fact that the set wheel/axles/chariot short circuits the rails.

In case of a bad shuntage, an additional device is used; its working, as well as its main parameters, is studied in this article.

A modeling process is applied with the support of the PEEC method to generate a complete electrical equivalent circuit of the device. Thanks to measurements, the accuracy of the modeling approach has been validated. The influence of geometrical and physical parameters on the performances of the studied device has been analyzed in order to find the main parameters and to optimize the structure.

Future works concerning the evaluation of the supplementary losses, the modeling of magnetic material as well as the modeling of the contact wheel/rail which is not fixed but flowing have to be achieved.

8. Acknowledgment

Authors want to thank the French National Railway Company for its financial support for this study and its help for the experimental sets.

9. References

Hoer, C. & Love, C. (1965). Exact Inductance Equations for Rectangular Conductors with Applications to More Complicated Geometries. *Journal of Research C. Engineering and Instrumentation*, Vol. 69C, No.2, (April-June 1965), pp. 127-137

Gonnet, J-P.; Clavel, E.; Mazauric; V. & Roudet, J. (2004). PEEC Method dedicated to the design of electrical systems. *Proceedings of PIERS 2004 Progress In Electromagnetic Research Symposium*, Pise, Italie, March 28-31 2004

Aimé, J.; Ardon, V.; Clavel; E., Roudet, J. & Loizelet, Ph. (2009a). EMC behavior of static converters thanks to radiated field modeling using an equivalent electrical circuit. *IEEE-IECON 35th International Annual Conference of the IEEE Industrial Electronics Society*, Porto, Portugal, November 2-5 2009

Ruehli, A.E. (1974). Equivalent circuit models for three dimensional multiconductor systems, *IEEE transaction on microwave theory and techniques*, Vol. MTT 22, No.3, (March 1974), pp. 216-221

Ardon, V.; Aimé, J.; Chadebec, O.; Clavel, E. & Vialardi, E. (2009). MoM and PEEC Method to Reach a Complete Equivalent Circuit of a Static Converter. *IEEE - EMC'09*, Zurich, Switzerland, January 12-16 2009

Clavel, E. & Prémont, Ch. (2007a). Function transfer sensitivity of an electronic filter versus capacitors location on a printed circuit board. *2EMC07* Rouen, France, October 18-19 2007

Jazzar, A.; Clavel, E.; Meunier, G.; Vincent, B.; Goleanu, A. & Vialardi, E. (2011). Modeling and simulating the lightning phenomenon: aeronautic materials comparison in conducted and radiated modes. *IEEE - ISIE11 20th International Symposium on Industrial Electronics*, Gdansk, Poland, June 27-30 2011

Ruehli, A. E. & Cangellaris, A. C. (2001). Progress in the methodologies for the electrical modeling of interconnects and electronic packages. *Proc. IEEE*, Vol. 89, no. 5, May 2001.

Kéradec, J-P.; Clavel, E.; Gonnet, J-P. & Mazauric, V. (2005). Introducing Linear Magnetic Materials in PEEC Simulations. Principles, Academic and Industrial Applications. *IEEE - IAS Industrial Applications Society*, Hong Kong, October 2-6 2005

Aimé, J.; Tran, S-T.; Clavel, E. & Meunier, G. (2009b). Far field extrapolation from near field interactions and shielding influence investigations based on a FE-PEEC coupling method. *IEEE Energy Conversion Congress and Exposition ECCE*, San José, Californie, USA, September 20-24 2009

Aimé, J.; Roudet, J.; Clavel, E.; Aouine, O.; Labarre, C.; Costa, F. & Ecrabey, J. (2007). Prediction and measurement of the magnetic near field of a static converter. *IEEE-ISIE 16th International Symposium on Industrial Electronics*, Vigo, Spain, June 4-7 2007, pp. 2550-2555.

Antonini, G.; Deschrijver, D. & Dhaene, T. (2007). Broadband Macromodels for Retarded Partial Element Equivalent Circuit (rPEEC) Method. *IEEE Transactions on Electromagnetic Compatibility*, Vol 49 , Issue: 1, 2007, pp. 35-48

Clavel, E.; Roudet, J.; Chevalier, Th. & Postariu, D. (2007b). Modelling connections taking into account return plane: application to EMI modelling for railway. . *IEEE-ISIE 16th International Symposium on Industrial Electronics*, Vigo, Spain, June 4-7 2007, pp. 2526-2531.

Vialardi, E.; Clavel, E.; Chadebec, O.; Guichon, J-M. & Lionet, M. (2010). Electromagnetic Simulation of Power Modules via Adapted Modelling Tools. *EPE-PEMC 14th International Power Electronics and Motion Control Conference*, Ohrid, Republic of Macedonia, September 6-8 2010.

Tan, S. X.-D.; He L. (2007) Advanced Model Order Reduction Techniques in VLSI Design, *Cambridge University Press*, New York, USA

## 6

Influence of the Phreatic Level on the Stability of Earth Embankments

Shodolapo Oluyemi Franklin[1] and Gbenga Matthew Ayininuola[2]

[1]*University of Botswana, Gaborone*
[2]*University of Ibadan, Ibadan*
[1]*Botswana*
[2]*Nigeria*

1. Introduction

Slopes in soils and rocks are common place in nature and man-made structures partly due to the fact that they are generally less expensive than constructing walls. However slope stability problems may arise due to the construction of artificial slopes in cuttings and embankments for roads and railways, or the construction of earth dams and water retaining embankments. Other reasons may include the study of the process of large scale natural slips or the application of remedial measures when such slips have taken place (Capper & Cassie, 1971). Existing slopes that have been stable may experience significant movement due to natural or man-made conditions. Such changes can result from the occurrence of earthquake, subsidence, erosion, the progression of tension or shrinkage cracks coupled with water ingress, changes in groundwater elevation or changes in the slope's subsurface flow which induces new seepage forces (McCarthy, 1998). Further causes may include the removal of earth below the toe of a slope or increased loading close to the crest of the slope. Slips may occur suddenly or gradually, commencing with a crack at the top of an earth embankment and slight upheaval near to the bottom and subsequently developing to a complete slip. All the foregoing actions make slip surface stability analysis of earth embankments complex and very difficult.

Several notable methods of analyzing slip surface stability have been developed over the years. Among the earliest was one that had its slip circle passing through soil materials whose shear strength is based upon internal friction and effective stresses (Fellenius, 1927). In this method an area of unit thickness of the volume tending to slide is divided into vertical strips and it is assumed that for each slice the resultant of the interslice forces is zero. A more significant and certainly more widely used approach utilizing the method of slices assumed a circular failure surface and fulfilled moment equilibrium but did not fully satisfy force equilibrium (Bishop, 1955). Yet another method considered a cylindrical slip surface and assumed that the forces on the sides of the slices are parallel (Spencer, 1967). A generalized approach (Morgenstern & Price, 1965) was developed in which all boundary and equilibrium equations are satisfied and the failure surface could assume any shape. The method involved solving systems of singular simultaneous equations and was unduly long in obtaining approximate answers despite the several assumptions made. An alternative generalized procedure (Bell, 1968) has been advanced which satisfies all conditions of equilibrium and assumed any failure surface. Here a solution is obtained by assuming a distribution of normal stress along the rupture surface. An approach involving the determination of the critical earthquake acceleration required to produce a condition of limiting equilibrium has been developed (Sarma, 1979). Also stability charts have been proposed which were partly based on the work of previous investigators and are applicable to a wide range of practical conditions (Cousins, 1978).

In practically all the afore-mentioned methods, the mass of soil assumed to be associated with the slope slide is divided into vertical slices. A slice is selected, the free body diagram of the forces acting on the slice is drawn and subsequently, based on limit equilibrium methods, an expression is derived for determining the factor of safety of the slope. The stability problem is dealt with by assuming that the tangential interslice forces are equal and opposite (Bishop, 1955). An iterative method for analyzing the stability problem in non-circular slip failures (Janbu, 1973) utilized a full rigorous algorithm and assumed a known line of thrust for the interslice horizontal forces. The method is best suited for computer solution. Similarly analytical solutions have been presented (Morgenstern & Price, 1965) which incorporate all interslice forces but are dependent on several assumptions and are quite lengthy.

1.1 Role of water pressure forces

The action of water is highly significant in slope movement. In clay and shale, softening by rain may lead to slip of a whole layer of material as a mud run. In addition water percolating into fissured clay may result in progressive deterioration and weakening that eventually results in reduction of shear strength so that a rotational or translational slip occurs. Consequently, for the various methods of slices highlighted above, it is important to stress the need to consider the water pressure forces acting not only at the interslice but also at the slice base, for such neglect may produce erroneous results.

More accurate but lower factors of safety are claimed for methods which account for the variation in seepage forces acting on and in the slice (King, 1989). Nevertheless such refinements depend on good estimates of water pressure. The factors of safety obtained for a given slope using different methods of slices are sensitive to the assumptions made in deriving them (Morrison & Greenwood, 1989). Furthermore the interslice forces play an important role in the resulting factors of safety. Hence the present study focuses on the effects of omitting the interslice pore water pressure forces on the overall stability of earth embankments and also endeavours to reduce the complexity features common to the more recent methods of slices outlined earlier.

2. Methodology

In general, any proposed method to evaluate the factor of safety must satisfy several requirements such as fulfilling limit equilibrium laws, account for all forces acting on the slice, adopt few assumptions, which should be easily comprehensible, be applicable to non-homogeneous soils, account for the water pressure distribution at the base of the slice as well as at the interslice, and treat stability problems in terms of both effective and total stresses. Furthermore, it is immaterial whether the horizontal interslice forces are considered as total or effective together with the force due to water pressure. When the force due to water pressure is correctly accounted for at the base as well as on the vertical sides of the slice, then the equilibrium of the body (soil skeleton) and water (hydrostatic) as well as normal and shear forces is maintained.

In order to properly assess the effect of water pressure forces on the stability of earth embankments, two algorithms which utilize both the limit equilibrium approach and method of slices are presented. The algorithms satisfy all the conditions stated above and in addition, the stability problem is treated as a 2-dimensional one in order to arrive at the final solution more quickly. A comprehensive description of the two algorithms and their application to a number of embankments reported in the literature is given elsewhere (Ayininuola & Franklin, 2008). Only the more important features will be considered here. The basic assumptions adopted for the present purpose are as follows: (a) The line of thrust on a slice and the pore water pressure forces are within the slice, preferably at one-third distance from the base or mid-point of the slice vertical sides (b) Factor of safety is defined in terms of the average shear stress developed along the potential failure surface and the average shear strength within the soil (c) Failure occurs simultaneously throughout the soil mass within the assumed rupture boundary.

2.1 Formulation A

Consider an elemental slice in Fig. 1. Due to the many unknown forces acting on the nth slice, the free body diagram in Fig. 2(a) is further divided into two separate bodies (Figs. 2(b) and 2(c)) under the basic requirements for the analysis of earth embankment stability. Since Fig. 2(a) is in equilibrium with all the forces acting on it, consequently Figs. 2(b) and 2(c) are in equilibrium as well.

On examining Fig. 2(a) critically, it is observed that of the twelve forces acting on the slice, only four have known magnitudes. This makes the analysis statically indeterminate of order eight. For the remaining eight forces to be determined there is need to establish logical relationships between the forces. With reference to Fig. 2(c) and letting $\Delta P_{w(n)}$ be the elemental increment of water pressure forces $P_{w(n+1)}$ and $P_{w(n)}$ across the slice, then

$$P_{w(n+1)} - P_{w(n)} = \Delta P_{w(n)} \tag{1}$$

The hydrostatic force U_n is obtained by measuring the free standing height of water in an installed piezometric tube at the slice base. If the piezometric height at the base is $H_{w(n)}$ then

$$U_n = \gamma_w \cdot H_{w(n)} \cdot g \tag{2}$$

where g is the acceleration due to gravity. Alternatively, since the forces acting in Fig. 2(a) are in equilibrium, it implies that the forces in Figs. 2(b) and 2(c) are also in equilibrium. Hence the hydrostatic force U_n at the base of the slice can be determined from the weight of water in the slice. From force equilibrium in the x and y directions the following expressions are obtained:

$$U_n = W_{w(n)} \sec \alpha_n \tag{3}$$

$$\Delta P_{w(n)} = -U_n \sin \alpha_n \tag{4}$$

where $W_{w(n)}$ is the weight of water in the nth slice and α_n is angle at the base of the slice. Substituting the value of U_n in equation (3) into equation (4) yields

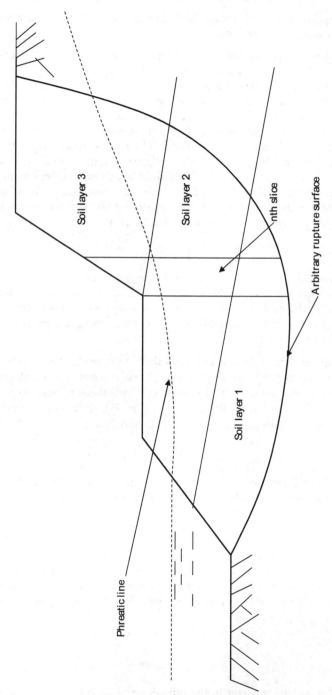

Fig. 1. Cross-section of an earth embankment made of non-homogeneous strata or soil layers

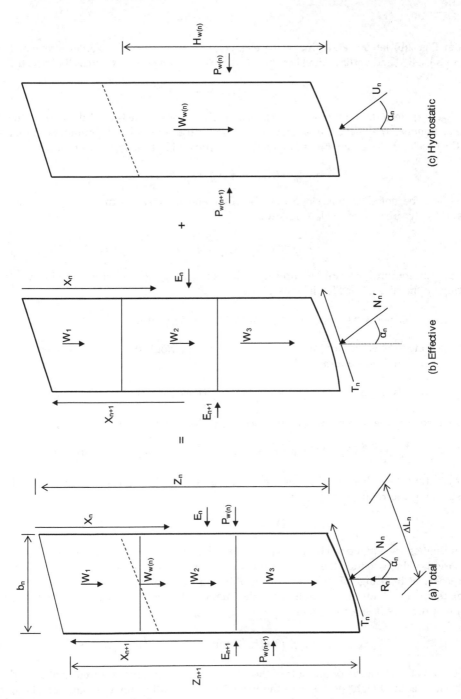

Fig. 2. Forces acting on an nth slice

$$\Delta P_{w(n)} = -W_{w(n)} \tan \alpha_n \tag{5}$$

From Fig. 2(b) let ΔE_n and ΔX_n be the elemental increments of horizontal interslice lateral thrusts E_n, E_{n+1} and vertical interslice shear forces X_n, X_{n+1} respectively across the slice, that is

$$\Delta E_n = E_n - E_{n+1}, \quad \Delta X_n = X_n - X_{n+1} \tag{6}$$

In Fig. 2(a) since the total normal force N_n and shear force T_n acting at the base of the slice are orthogonal, they have the same resultant R_n. In order to reduce the number of unknowns at the base of the slice the forces T_n and N_n are expressed in terms of R_n as follows:

$$T_n = R_n \sin \alpha_n, \quad N_n = R_n \cos \alpha_n \tag{7}$$

It should be noted at this stage that the total and effective normal forces, N_n and N_n' respectively, are related to U_n as follows:

$$N_n' = N_n - U_n \tag{8}$$

Limit equilibrium laws can be applied to the slice in Fig. 2(a). Firstly resolving the forces acting on the slice in the N_n-direction yields

$$N_n - W_n \cos \alpha_n - \Delta X_n \cos \alpha_n + \Delta E_n \sin \alpha_n + \Delta P_{w(n)} \sin \alpha_n - W_{w(n)} \cos \alpha_n = 0 \tag{9}$$

Substituting the values of $\Delta P_{w(n)}$ and N_n from equations (5) and (7) respectively into equation (9) and re-arranging yields

$$R_n \cos \alpha_n + \Delta E_n \sin \alpha_n - \Delta X_n \cos \alpha_n = W_{w(n)} \left(\cos \alpha_n + \tan \alpha_n \sin \alpha_n \right) + W_n \cos \alpha_n \tag{10}$$

Also, resolving the forces acting on the slice in the T_n-direction yields

$$T_n - W_n \sin \alpha_n - \Delta X_n \sin \alpha_n - \Delta E_n \cos \alpha_n - W_{w(n)} \sin \alpha_n - \Delta P_{w(n)} \cos \alpha_n = 0 \tag{11}$$

Substituting the values of $\Delta P_{w(n)}$ and T_n from equations (5) and (7) into equation (11) and simplifying and re-arranging results in

$$R_n \sin \alpha_n - \Delta X_n \sin \alpha_n - \Delta E_n \cos \alpha_n = W_n \sin \alpha_n \tag{12}$$

Examination of equations (10) and (12) reveals that there are three unknowns in the two equations which render them indeterminate; in order to solve for these unknowns it is necessary at this stage to introduce an equilibrium equation based on moments. From Fig. 2(b), taking moments of the resultants of the interslice forces and other forces about the mid-point of the slice base width gives

$$\Delta X_n . b_n / 2 - \Delta E_n \left\{ Z/3 + \left(b_n/2 \right) \tan \alpha_n \right\} - \Delta P_{w(n)} \left(H_{w(n)} / 3 \right) = 0 \tag{13}$$

where Z is the elevation of one side of the slice, or more correctly, the greater of Z_n and Z_{n+1}, and $H_{w(n)}$ is the height of water in the nth slice. Substituting for the value of $\Delta P_{w(n)}$ from equation (5) into equation (13) and re-arranging results in

$$-\Delta E_n\{Z/3+(b_n/2)\tan\alpha_n\}+\Delta X_n.b_n/2=-W_{w(n)}\tan\alpha_n\left(H_{w(n)}/3\right) \tag{14}$$

Equations (10), (12) and (14) can now be assembled together as a set of simultaneous equations in the following form

$$\left.\begin{array}{l} R_n\cos\alpha_n+\Delta E_n\sin\alpha_n-\Delta X_n\cos\alpha_n=W_{w(n)}\left(\cos\alpha_n+\tan\alpha_n\sin\alpha_n\right)+W_n\cos\alpha_n \\[4pt] R_n\sin\alpha_n-\Delta X_n\sin\alpha_n-\Delta E_n\cos\alpha_n=W_n\sin\alpha_n \\[4pt] -\Delta E_n\{Z/3+(b_n/2)\tan\alpha_n\}+\Delta X_n.b_n/2=-W_{w(n)}\tan\alpha_n\left(H_{w(n)}/3\right) \end{array}\right] \tag{15}$$

In matrix format the simultaneous equation (15) becomes

$$\begin{bmatrix} \cos\alpha_n & \sin\alpha_n & -\cos\alpha_n \\ \sin\alpha_n & -\cos\alpha_n & -\sin\alpha_n \\ 0 & -\{Z/3+(b_n/2)\tan\alpha_n\} & b_n/2 \end{bmatrix} \begin{bmatrix} R_n \\ \Delta E_n \\ \Delta X_n \end{bmatrix}$$
$$= \begin{bmatrix} W_{w(n)}(\cos\alpha_n+\tan\alpha_n\sin\alpha_n)+W_n\cos\alpha_n \\ W_n\sin\alpha_n \\ -W_{w(n)}\tan\alpha_n(H_{w(n)}/3) \end{bmatrix} \tag{16}$$

In compact form this can be written as

$$K.D=F \tag{17}$$

$$\text{where } K=\begin{bmatrix} \cos\alpha_n & \sin\alpha_n & -\cos\alpha_n \\ \sin\alpha_n & -\cos\alpha_n & -\sin\alpha_n \\ 0 & -\{Z/3+(b_n/2)\tan\alpha_n\} & b_n/2 \end{bmatrix}, \quad D=\begin{bmatrix} R_n & \Delta E_n & \Delta X_n \end{bmatrix}^{-1}$$

$$\text{and } F=\begin{bmatrix} W_{w(n)}(\cos\alpha_n+\tan\alpha_n\sin\alpha_n)+W_n\cos\alpha_n & W_n\sin\alpha_n & -W_{w(n)}\tan\alpha_n\left(H_{w(n)}/3\right) \end{bmatrix}^{-1}$$

In the above expressions the matrices D and F represent the nodal unknown forces and the nodal applied forces respectively.

Equation (17) above is for the nth slice, and when such sets of equations are assembled for the whole rupture mass this yields

$$K_g.D_g=F_g \tag{18}$$

where K_g is the global stiffness matrix, D_g is the global forces matrix and F_g is the global applied forces matrix. The simultaneous equation (18) can be solved for various values of R_n, ΔE_n and ΔX_n using the Gaussian elimination, Jacobi or Gauss-Siedel iterative techniques. The Gaussian elimination method is, relatively speaking, the simplest and easiest of the three procedures to implement. Once the values of R_n are obtained, the values of N_n and T_n can also be found using equations (7) and (8).

The factor of safety can be defined in terms of the shear strength of the soil and the shear stress developed along the potential failure surface based on the Coulomb-Mohr failure criteria in terms of effective stress as follows:

$$T_n = \left(c_n' \Delta L_n + N_n' \tan\phi_n' \right) / F_s \tag{19}$$

where F_s is the factor of safety and ϕ_n' is the angle of shearing resistance with respect to effective stress. Substituting for the values of T_n and N_n' from equations (7) and (8) into equation (19) yields

$$R_n \sin\alpha_n = \left[c_n' \Delta L_n + \left(R_n \cos\alpha_n - U_n \right) \tan\phi_n' \right] / F_s \tag{20}$$

Also substituting the value of U_n from equation (3) into equation (20) and then considering the whole of the rupture mass consisting of the set of slices will give

$$\sum R_n \sin\alpha_n = \sum \left[c_n' \Delta L_n + \left(R_n \cos\alpha_n - W_{w(n)} \sec\alpha_n \right) \tan\phi_n' \right] / F_s \tag{21}$$

Making F_s the subject of the expression in equation (21) will yield

$$F_s = \frac{\sum \left[c_n' \Delta L_n + \left(R_n \cos\alpha_n - W_{w(n)} \sec\alpha_n \right) \tan\phi_n' \right]}{\sum \left(R_n \sin\alpha_n \right)} \tag{22}$$

Equation (22) can be use to analyze stability problems involving both homogeneous and non-homogeneous soils types.

Irrespective of whether the earth embankment is partially or wholly drained, the equation can be applied because during its formulation both states of stress were taken into account.

2.2 Formulation B

The present study seeks to investigate the effect of hydrostatic pore water pressure forces on the overall stability of earth embankments and as such, in order to establish a basis of comparison with the earlier algorithm presented, an alternative approach is developed. This treats stability problems in terms of effective stresses and assumes that the influence of water pressure forces acting at the interslice can be neglected. The lines of action of $P_{w(n+1)}$ and $P_{w(n)}$ are taken to be coincident and also $\Delta P_{w(n)} = 0$. Proceeding along the same lines as the previous formulation, the following set of simultaneous equations can be arrived at:

$$\left.\begin{array}{l} R_n \cos\alpha_n + \Delta E_n \sin\alpha_n - \Delta X_n \cos\alpha_n = \left(W_{w(n)} + W_n \right) \cos\alpha_n \\[2mm] R_n \sin\alpha_n - \Delta X_n \sin\alpha_n - \Delta E_n \cos\alpha_n = \left(W_{w(n)} + W_n \right) \sin\alpha_n \\[2mm] -\Delta E_n \left\{ Z/3 + \left(b_n/2 \right) \tan\alpha_n \right\} + \Delta X_n . b_n/2 = 0 \end{array}\right] \tag{23}$$

For the typical nth slice the above equation in matrix format becomes

$$
\begin{bmatrix}
\cos\alpha_n & \sin\alpha_n & -\cos\alpha_n \\
\sin\alpha_n & -\cos\alpha_n & -\sin\alpha_n \\
0 & -\{Z/3+(b_n/2)\tan\alpha_n\} & b_n/2
\end{bmatrix}
\begin{bmatrix}
R_n \\
\Delta E_n \\
\Delta X_n
\end{bmatrix}
=
\begin{bmatrix}
(W_{w(n)}+W_n)\cos\alpha_n \\
(W_{w(n)}+W_n)\sin\alpha_n \\
0
\end{bmatrix}
\tag{24}
$$

Again proceeding along the same lines as the previous formulation, an expression very similar to equation (22) can be obtained as follows:

$$
F_s = \frac{\sum\left[c_n'\Delta L_n + \left(R_n\cos\alpha_n - W_{w(n)}\sec\alpha_n\right)\tan\phi_n'\right]}{\sum(R_n\sin\alpha_n)}
\tag{25}
$$

Although equations (25) and (22) are very similar, the procedures for evaluating the values of R_n in both equations are certainly not the same. Consequently different values of factors of safety will be obtained using both approaches. The methods developed can be used for slip surface stability analysis either manually or with a programmable calculator. However while this may be true for fairly homogeneous slopes, for real or non-homogeneous soils the computation work is quite daunting for practical design. This is on account of the number of rupture surfaces that may need to be analyzed in order to obtain the most critical rupture surface for design purposes as well as the fact that the global stiffness matrix K_g mentioned earlier may be of the order 60 x 60 or more, depending on the number of slices within the rupture mass. Hence comprehensive computer software was developed involving two minimization computer programmes which can handle problems of up to three soil strata; some details of the programmes are given elsewhere (Ayininuola, 1999).

3. Results

In order to assess the effect of the pore water pressure forces, the procedures developed in the present study have been applied to a number of earth embankments some of which are reported in the literature. Firstly, the Lodalen Landslide (Sevaldson, 1956) is examined and then, the case of a non-homogeneous earth dam (Sherard et al, 1978) is investigated. Finally the effect of altering the phreatic level on the Okuku dam in South-Western Nigeria (Ayininuola & Franklin, 2008) is studied.

3.1 Stability analysis of the Lodalen Landslide (Sevaldson, 1956)

Fig. 3 shows a sectional view of the Lodalen Landslide. A stability analysis of the earth embankment prior to the occurrence of the slide will be carried out. Towards this end the initial rupture surface has been divided into 13 slices. The necessary data have been taken from the initial rupture surface and fed into the computer programme mentioned earlier. A total of 100 rupture surfaces have been considered in the analysis. Details of the computer output are not presented here, but a summary of the main findings are shown in Table 1 and these results are discussed at a later stage.

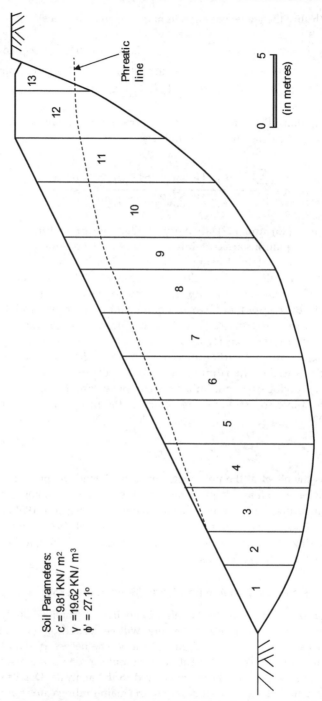

Fig. 3. Re-examination of the Lodalen slide (Modified from Sevaldson, 1956)

3.2 Stability analysis of a non-homogeneous earth dam (Sherard et al, 1978)

A non-homogeneous earth dam is shown in section in Figs. 4 and 5 and it is required to carry out a stability analysis of both the upstream and downstream sides of the dam. The assumed initial rupture surface on both sides of the dam has been divided into 12 slices each and the necessary data taken from the rupture surfaces have been fed into the computer programme referred to earlier. A total of 200 rupture surfaces have been considered at both the upstream and downstream sections. Again details of the computer output are not given in the present study, however a summary of the main findings are shown in Table1.

Earth Embankment	Method	Factor of Safety
Lodalen slide (Sevaldson, 1956)	Authors' Formulation A Authors' Formulation B Bishop's Simplified	0.80 0.90 0.90
Downstream section of non-homogeneous earth dam (Sherard et al, 1978)	Authors' Formulation A Authors' Formulation B Bishop's Simplified	1.49 1.83 1.83
Upstream section of non-homogeneous earth dam (Sherard et al, 1978)	Authors' Formulation A Authors' Formulation B Bishop's Simplified	0.66 0.78 0.78

Table 1. Results of stability analysis of Lodalen slide and a non-homogeneous earth dam

3.3 Stability analysis of Okuku earth dam, Nigeria

On account of the accessibility to data, the Okuku dam has been utilized as a case study in order to investigate and understand the response of the proposed formulations to changes in the phreatic levels in the earth embankment due to variation in water levels in the storage reservoir. The dam was constructed in 1995 at Okuku town located on the 8° 02'N and 4° 40'E coordinates and approximately 40 km North-East of Osogbo in Osun State, South-Western Nigeria. The dam axis located across River Anle, a seasonal stream, is about 1.5 km South-East of Okuku town. The dam is a homogeneous earth dam built with poorly graded sand clay mixtures which possess the following soil characteristics, namely, cohesion c' = 45 KN/m^2, angle of shearing resistance ϕ' = 12° and additionally, average dry density of dam construction materials, γ = 19.63 KN/m^3. The height of the crest above the base of the dam is 10 metres and the upstream and downstream sections are sloped at ratios 1:3 and 1:2.5 respectively. In Figs. 6 and 7, diagrams of the dam embankment for both the upstream and downstream sections at different levels of water in the storage reservoir are shown. Additional details in respect of the dam design may be found elsewhere (Ayininuola, 1999). The factors of safety at different phreatic levels for both the upstream and downstream sections of the dam have been estimated using the two formulations developed earlier. A total of 500 rupture surfaces have been examined for each section. A summary of the results is presented in Table 2 and Figs. 8 and 9.

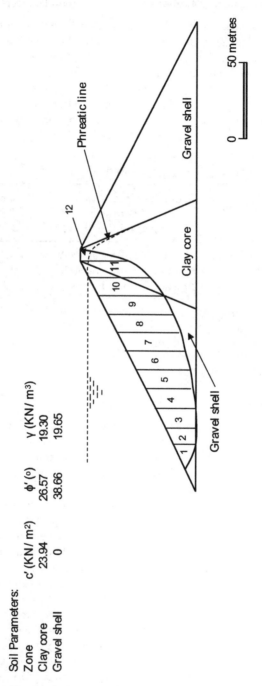

Fig. 4. Upstream section of an earth dam embankment (Modified from Sherard et al, 1978)

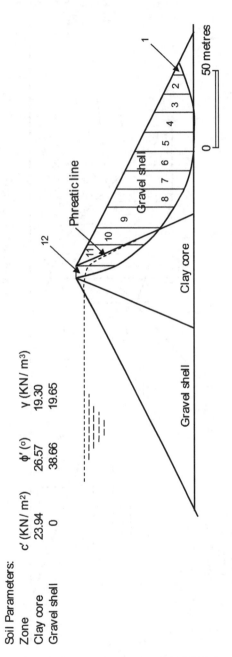

Fig. 5. Downstream section of an earth dam embankment (Modified from Sherard et al, 1978)

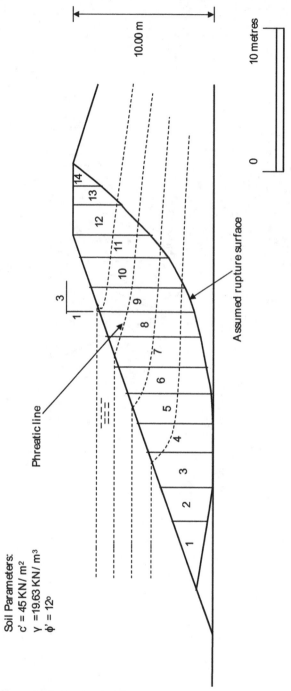

Fig. 6. Upstream section of Okuku earth dam (Courtesy Konsadem Associates Ltd., Nigeria)

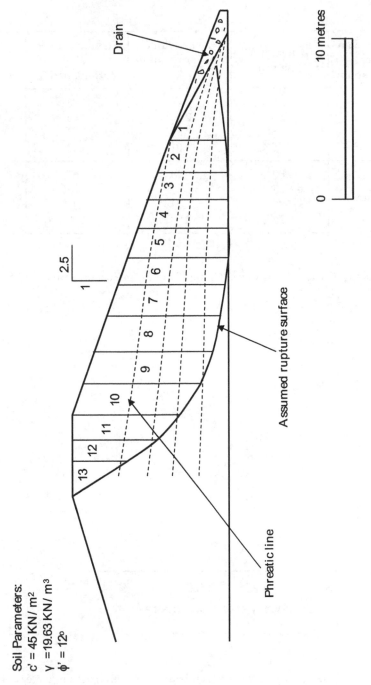

Fig. 7. Downstream section of Okuku earth dam (Courtesy Konsadem Associates Ltd., Nigeria)

Section of dam under consideration	Water level (metres)	Stability values for Formulation A	Stability values using formulation B	Difference (%)
Upstream	9.00*	2.43	2.78	14.40
	6.55*	2.60	2.85	9.62
	5.25*	2.81	2.93	4.27
	3.70*	3.01	3.03	0.66
Downstream	9.00**	2.25	2.67	18.67
	6.55**	2.53	2.76	9.09
	5.25**	2.74	2.83	3.28
	3.70**	2.89	2.90	0.35

Table 2. Results of stability analysis of Okuku Dam using Formulation A and B

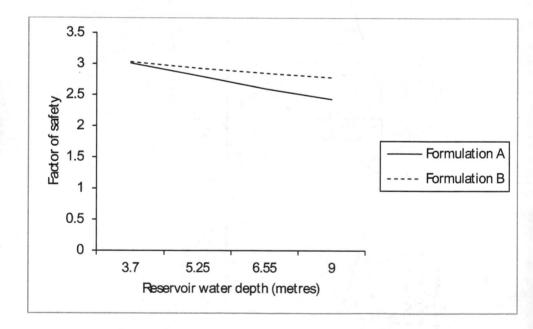

Fig. 8. Variation of factor of safety of dam embankment with reservoir water depth (upstream section)

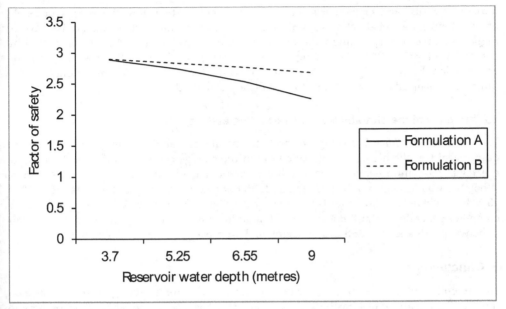

Fig. 9. Variation of factor of safety of dam embankment with reservoir water depth (downstream section)

4. Discussion

The comments outlined here are based primarily on the results presented in Tables 1 and 2, as well as in Figs. 8 and 9. As noted in the preceding section, details of the computer output in respect of the stability analysis carried out are not presented here, but may be obtained elsewhere (Ayininuola & Franklin, 2008; Ayininuola, 1999).

4.1 Comparison between formulations A and B and Bishop's simplified method

The procedures developed in the present study as well as Bishop's have been applied to the Lodalen landslide as well as the non-homogeneous earth dam described in the preceding section. Using Table 1 and the results of the computer generated output for the Lodalen landslide as guide, several points may be noted. Firstly the water pressure forces acting at the interslice have great influence on the elemental horizontal thrusts generated at the interslice. They also have a direct influence on the elemental shear forces at the interslice. In this region when the elemental water pressure forces are assumed to be zero, the values of the elemental horizontal thrusts and shear forces that develop are much smaller than those obtained when water pressure forces are taken into account.

In addition to the above, water pressure forces, elemental horizontal thrusts and elemental shear forces are directly affected by the slice inclination angles. When for example the inclination angle is zero, the effect of all the forces mentioned is practically negligible. At the interslice when the piezometric height $H_{w(n)}$ is zero, the values of elemental horizontal thrust and elemental shear force are zero. Furthermore the factors of safety obtained when the net effect of water pressure forces at the interslice is taken to be zero are higher than the corresponding values when the afore-mentioned forces are considered by the order of 8% – 24% depending on the phreatic level. Also the results given by Formulation B, which ignores the water pressure forces effect at the interslices, are practically identical to those obtained using Bishop's method.

4.2 Influence of the phreatic level on earth dam stability

With reference to Table 2 and Figs. 8 and 9, the results of stability analysis of Okuku earth dam reveal that the higher the phreatic level in the storage reservoir of the earth dam, the greater the variation between the stability values obtained using Formulations A and B. When the water pressure forces are ignored, higher factors of safety of the order of 0.35% – 18.67% are obtained. With regards to the curves in Figs. 8 and 9, and also from a study of the computer generated output, it is observed that at low phreatic levels in the storage reservoir of between 20% and 30% of the dam height, both approaches yield similar stability values.

5. Conclusions

An in-depth study of the effect of the pore water pressure forces acting at the interslice on the stability of earth dams has been carried out. This has been achieved by developing two procedures and applying the formulations to a number of practical cases. Based on the results of the investigation, a number of conclusions can be drawn: Firstly the magnitudes of effective horizontal thrusts and shear forces generated at the interslice when pore water pressure forces induced are taken into consideration are higher than those obtained when these forces are ignored. This demonstrates that the pore water pressures developed have an influence on the values of other interslice forces. Secondly the inclusion of net water pressure forces in the stability analysis of the earth embankments studied clearly show that the action of the water pressure forces serves to promote instability, as would be expected. Thirdly the popular practice amongst geotechnical engineers of resolving the water pressure within a given slice in a direction of normal at the slice base in order to estimate its value, whilst the horizontal effect of the slice base water pressure is taken as zero, constitutes a grave error. This action is not in line with limit equilibrium procedures and yields erroneous results. Fourthly at low phreatic levels the proposed approaches give practically similar factors of safety. This implies that the effect of water pressure forces acting on the interslice can only be neglected when the phreatic line in an embankment is at its lowest stage, a considerable period after drawdown, or preferably between 20% and 30% of the overall height of the embankment. Finally the factors of safety found using Bishop's simplified method and that based on Formulation B, which ignores the pore water pressure forces effect, are in very close agreement. This simply implies that the inclusion of only the effective horizontal thrusts and shear forces acting at the interslice has little influence on the resulting factors of safety.

6. Acknowledgments

The present investigators would wish to thank the management of Konsadem Associates Limited, Ibadan, Nigeria for making readily available the data and other aspects relating to the design of the Okuku earth dam embankment. In addition the assistance received from staff of Ete-Aro and Partners Limited, Ibadan, is gratefully acknowledged. The authors would also wish to place on record the encouragement of staff of the Department of Civil Engineering, University of Ibadan, Nigeria as well as staff of the Department of Civil Engineering, University of Botswana, Gaborone, Botswana.

7. References

Capper, P. & Cassie, W. (1971). *The Mechanics of Engineering Soils,* 5th Edition, E. & F.N. Spon, ISBN 419-10700-2, London, United Kingdom

McCarthy, D. (1998). *Essentials of Soil Mechanics and Foundations - Basic Geotechnics,* 5th Edition, Prentice-Hall, ISBN 0-13-506932-7, New Jersey, USA

Fellenius, W. (1927). *Erdstatische Berechnungen mit Reibung und Kohasion Adhasion und unter Annahme Kreiszylindrischer Gleitflachen,* W. Ernst, Berlin

Bishop, A. (1955). The Use of the Slip Circle in the Stability Analysis of Slopes, *Geotechnique,* Vol.5, No.1, (March 1955), pp 7–17, ISSN 0016-8505

Spencer, E. (1967). A Method of Analysis of the Stability of Embankments Assuming Parallel Interslice Forces, *Geotechnique,* Vol.17, No.1, (March 1967), pp 11–26, ISSN 0016-8505

Morgenstern, N. & Price, V. (1965). The Analysis of the Stability of General Slip Surfaces, *Geotechnique,* Vol.15, No.1, (March 1965), pp 79–93, ISSN 0016-8505

Bell, J. (1968). General Slope Stability Analysis, *J. Soil Mech. & Found. Div., ASCE,* Vol.94, No.6, (June 1968), pp. 253–270, ISSN 04447994

Sarma, S. (1973). Stability Analysis of Embankments and Slopes, *Geotechnique,* Vol.23, No.2, (June 1973), pp. 423–433, ISSN 0016-8505

Cousins, B. (1978). Stability Charts for Simple Earth Slopes, *J. Geotechnical. Eng. Div., ASCE,* Vol.104, No. GT2, (February 1978), pp. 267–279, ISSN 00936405

Janbu, N. (1973). Slope Stability Computations, In: *Embankment Dam Engineering- Casagrande Memorial Volume,* R.C. Hirschfield & S.J. Poulos (Ed.), 47–86, John Wiley & Sons, ISBN 0-471-40050-5, New York, USA

King, G. (1989). Revision of Effective Stress Method of Slices, *Geotechnique,* Vol.39, No.3, (September 1989), pp. 497–502, ISSN 0016-8505

Morrison, I. & Greenwood, J. (1989). Assumptions in Simplified Slope Analysis by the Method of Slices, *Geotechnique,* Vol.39, No.3, (September 1989), pp. 503–509, ISSN 0016-8505

Ayininuola, G. & Franklin, S. (2008). Water Pressure Forces Effect on Earth Embankments Stability, *Global Journal of Engineering and Technology,* Global Research Publication, Vol.1, No.2, (June 2008), pp. 169–188

Ayininuola, G. {1999). *The Effect of Hydrostatic Pore Water Pressure Forces on the Stability of Earth Embankments,* Unpublished M.Sc. Dissertation, University of Ibadan, Nigeria.

Sevaldson, R. (1956). The Slide in Lodalen, Oct. 6th 1954, *Geotechnique,* Vol.6, No.4, (December 1956), pp 167–191, ISSN 0016-8505

Sherard, J., Woodward, R., Gizienski, S. & Clevenger, W. (1978). *Earth-Rock Dams: Engineering Problems of Design and Construction*, J. Wiley and Sons, ISBN 0-471-78547-4, New York, USA

## 7

Special Tunnel Blasting Techniques for Railway Projects

More Ramulu

Central Institute of Mining & Fuel Research, Nagpur,
India

1. Introduction

Excavations of tunnels are common features in mining and civil engineering projects. In absence of initial free face, solid blasting method is employed for excavation of tunnels, drifts and mine roadways, which have many similarities in configurations and in different cycles of operation followed during excavation. A greater proportion of world's annual tunnel advance is still achieved by drilling and blasting. In spite of inherent disadvantages of damaging the rock mass, drilling and blasting has an unmatched degree of flexibility and can overcome the limitations of machine excavations by Tunnel Boring Machine (TBM) or road headers. In spite of no major technical breakthrough, the advantages like low investment, availability of cheap chemical energy in the form of explosives, easy acceptability to the practicing engineers, the least depreciation and wide versatility have collectively made the drilling and blasting technique prevail so far over the mechanical excavation methods.

Since tunnels of different sizes and shapes are excavated in various rock mass conditions, appropriate blast design including drilling pattern, quantity and type of explosive, initiation sequence is essential to achieve a good advance rate causing minimal damage to the surrounding rock mass. The cost and time benefit of the excavation are mostly decided by the rate of advance and undesired damage.

Excavation of tunnels, except in geologically disturbed rock mass conditions, is preferred with full face blasting. It is common to excavate large tunnels of 80-90 m² cross-section in sound rock masses by full face in a single round. However, tunnels larger than 50m² cross-sectional area driven through incompetent ground condition are generally excavated in smaller parts.

Introduction of electro-hydraulic jumbo drills with multiple booms, non-electric initiation system, small diameter explosives for contour blasting and fracture control blasting are some of the recent developments in tunnel blasting. Prediction and monitoring the blast damage, application of computers in drilling, numerical modelling for advanced blast design, use of rock engineering systems for optimization and scheduling of activities have been the areas of intense research in today's competitive and high-tech tunnelling world

In tunnel blasting, explosives are required to perform in a difficult condition, as single free face (in the form of tunnel face) is available in contrast to bench blasting where at least two free faces exist. Hence, more drilling and explosives are required per unit volume of rock to be fragmented in the case of tunnel blasting. A second free face, called 'cut', is created initially during the blasting process and the efficiency of tunnel blast performance largely depends on the proper development of the cut. The factors influencing the development of the cut and the overall blast results are dependent on a host of factors involving rock mass type, blast pattern and the tunnel configurations.

2. Blasting mechanics

The tunnel blasting mechanics can be conceptualised in two stages. Initially, a few holes called cut holes are blasted to develop a free face or void or cut along the tunnel axis. This represents a solid blasting condition where no initial free face is available. Once the cut is created, the remaining holes are blasted towards the cut. This stage of blasting is similar to bench blasting but with larger confinement. The results of tunnel blasting depend primarily on the efficiency of the cut hole blasting. The first charge fired in cut resembles crater blasting. Livingston's spherical charge crater theory (Livingston, 1956) suggests that the blast induced fracturing is dominated by explosion gas pressure which is supported by Liu and Katsabanis (1998). Duvall and Atchison (1957), Wilson (1987) and others believe that the stress wave induced radial fracturing is the dominating cause of blast fragmentation and gas pressure is responsible only for extension of the fractures developed by the stress wave.

The natures of influence of the two pressures i.e. of stress and gas are different in the jointed rock mass where the stress waves is useful in fragmentation as the joints restrict the stress wave propagation. The gases, on the other hand, penetrate the joint planes and try to separate the rock blocks. The fragments' size and shape in jointed formations are dominated by the gas pressure and the joint characteristics. The roles of the stress wave and the gas pressures are no different in the second stage of tunnel blasting. But with the availability of free face, the utilisation of stress wave is increased. The rock breakages by rupturing and by reflected tensile stress are more active in the second stage because of cut formation in the first stage.

3. Parameters influencing tunnel blast results

The parameters influencing the tunnel blast results may be classified in three groups:

i.	Non controllable	Rock mass properties,
ii.	Semi-controllable	(a) Tunnel geometry & (b) Operating factors, and
iii.	Controllable	Blast design parameters including the explosive properties.

4. Models for prediction of tunnel blast results

Specific Charge is one of the important paprameter of prediction of tunnel blast results. Pokrovsky (1980) suggested the following empirical relation to determine the specific charge (q) in tunnels (Eq. 1):

$$q = q1. \, s_t. \, f. \, s_{wr}. \, d_{ef}, \, kg/m^3 \qquad (1)$$

where,

$q1$=specific charge for breaking of rock against a free face in kg/m^3,
s_t = factor for structure and texture of rock,

$$f = rock \, confinement = 6.5 \, / \, \sqrt{A}, \qquad (1a)$$

A= area of tunnel (m^2),
s_{wr} = relative weight strength of explosive (ANFO = 1), and
d_{ef} = factor for diameter of explosive cartridge,

According to Langefors and Kihlstrom (1973), the specific charge (q) is related to the cross-sectional area of the tunnel (A, m^2) as given below:

$$q = (14/A) + 0.8 \, kg/m^3 \qquad (2)$$

The specific charge in the cut holes remain maximum and it can be upto 7 kg/m^3 in a parallel cut.

5. Rock mass damage

The aspects of blast induced rock mass damage around a tunnel opening and its assessment have been the subjects of in-depth research for quite a long time. The type of damage can be grouped into three categories: (i) fabric damage due to fracturing, (ii) structural damage exploiting discontinuities and shears, and (iii) lithological damage causing parting between two different rock units or lithological boundaries between similar rock types.

Chakraborty et al. (1996a) observed in the tunnels of Koyna Hydro-electric Project, Stage–IV poor pull and small overbreak in volcanic breccia having low Q value, P-wave velocity and modulus of elasticity. On the other hand, large overbreak on the sides due to vertical and sub-vertical joints and satisfactory pull were found in the compact basalts having comparatively much higher Q value, P-wave velocity and modulus of elasticity. The fact is attributed to the presence of well defined joints in compact basalts which is absent in volcanic breccia.

The effects of joint orientations on overbreak/underbreak and pull in heading and benching operations during tunnel excavations are explained by Johansen (1998). The work of Johansen (1998) describes that joints normal to tunnel direction are favorable for good pull and parallel to the tunnel advance direction yield poor pull. advance direction. The obtuse angle between joints and tunnel direction results in more damage and breakage towards the wall of that angle.

The dip direction of the blasted strata on pull could be well experienced while blasting in the development faces of Saoner coal mine where the pull was increased by 11 per cent in the rise galleries compared to that in the dip galleries (Chakraborty, 2002). Longer rounds in tunnels can be pulled when the dominant joint sets are normal to the tunnel axis. Whereas, better pull can be obtained in shaft sinking if the discontinuities are parallel to the line joining the apex of the Vs in a V-Cut Hagan (1984).

Chakraborty (2002) observed the following influences of joint directions on pull and overbreak (Table 1).

Joint Orientation		Face Advance	Roof Overbreak
Dip	Strike with respect to tunnel axis		
Steep	Parallel	Very poor	Very small
Steep	Across	Very good	Very large
Gentle	Across	Fair	Large
Moderate	Across/oblique	Good	Small

Table 1. Influence of joint direction on overbreak (Chakraborty, 2002)

The gentle, moderate and steeply dipping joint planes signify the dip angles as 0°-30°, 30°-60° and 60°-90° respectively. Similarly, strikes with respect to tunnel axis are mentioned as parallel, oblique and across to indicate that the joint strike intersection angle with the tunnel axis as 0°-30°, 30°-60° and 60°-90° respectively.

If the geo-mechanical properties of the constituting formations of a tunnel are quite different, the stress energy utilisation and resulting fragmentation are adversely affected. Chakraborty et al. (1996b) suggested an increase of specific charge by a per cent equal to ten times the number of contact surfaces.

Engineers International Inc. modified Basic RMR (MBR) considering blast-induced-damage adjustments, as shown in Table 2, were suggested for planning of caving mine drift supports (Bieniawski,1984). Chapter 4 in the present publication defines basic RMR.

Method of Excavation	Damage Level	Blast Damage Adjustment Factor	Per cent Reduction
1. Machine boring	No damage	1.0	0
2. Controlled blasting	Slight	0.94-0.97	3-6
3. Good conventional blasting	Moderate	0.9-0.94	6-10
4. Poor conventional blasting	Severe damage	0.9-0.8	10-20

Table 2. Blast damage adjustments in MBR (after Bieniawski, 1984)

Ouchterlony et al. (1991) observed that the damage zone could be to the extent of 0.5 m with cautious tunnel blasting. McKenzie (1994) related the threshold peak particle velocity PPV (v_{max}) for incipient fracture with uniaxial tensile strength (q_t), Young's modulus and P-wave velocity (V_p, m/s) as shown below:

$$v_{max} = \frac{q_t \times V_p \times 10^{-3}}{E} , m/s \tag{3}$$

where

q_t = uniaxial tensile strength, MPa,
V_p = P-wave velocity, m/s, and
E = Young's modulus, GPa.

Pusch and Stanfors (1992) and others observed that the minimum disturbance by blasting is reported when the tunnel orientation was within 15° with the strike of the joint sets.

Yu and Vongpaisal (1996) concluded that the damage is a function of blast induced stress and rock mass resistance to damage. They proposed Blast Damage Index (D_{ib}) to estimate the type of damage due to blasting. It is the ratio of the blast induced stress to the resistance offered against damage.

Ramulu et al (2009) categorised blast induced damage as,

- near-field damage due to high frequency and critical vibrations
- far-field damage due to repeated low frequency and sub-critical vibrations. Ramulu and Sitharam (2011) assessed the near-filed damage by using vibration attenuation model of charge weight scaling law and dynamic tensile failure criteria instead of conventional Holmberg-Persson model (1979) and static tensile failure criteria. Ramulu (2010) correlated the far-filed damage with shear wave velocity of rock mass and found the following equation with reasonably good correlation coefficient (R^2=0.76).

$$D_{max}=322.5(V_s)^{-0.61} \text{ m} \tag{4}$$

where,

D_{max} – Maximum extent of rock mass damage due to repeated vibrations, m
V_s – S-wave velocity, m/s

6. Contour blasting

Contour blasting in tunnelling is adopted to obtain a smooth tunnel profile and minimise damage to the surrounding rock mass. Despite a large amount of drilling required, it is preferred over conventional blasting because of the following advantages:

i.   The shape of the opening is maintained with smooth profile.
ii. Stability of the opening and the stand-up-time of the tunnel are improved.
iii. Support requirement is reduced.
iv. Overbreak is reduced to minimise unwanted excavations and filling to bring down the cost and cycle time.
v.   Ventilation improves due to smooth profile.

The performance of contour blasting is frequently measured in terms of `Half cast factor' (HCF) which is dominated by the design parameters of the contour holes, the joint orientation and the explosive energy distribution.

Generally, two types of contour blasting are used in tunnelling, i) pre-splitting and ii) smooth blasting. When two closely spaced charged holes are fired simultaneously the stress waves generated from the two holes collide at a plane in between the holes and create a secondary tensile stress front perpendicular to the hole axis and facilitates extension of radial cracks along the line joining the holes. The wedging action from the explosion gas acts in favour of extending the crack along the same line. It is, therefore, essential to contain the gas pressure till the cracks from both ends meet by adequate stemming. Further, the delay timing of the adjacent holes need to be very accurate so that the stress waves should collide at the mid-point and the arbitrariness of the breakage between the holes can be reduced.

The contour blasting performance largely depends on the nature and the orientation of joint planes. Gupta et al. (1988) found that the joint orientation adversely influences the pre-splitting results to a maximum when these are at an angle of 1-30° to the pre-split axis.

In smooth blasting, the delay intervals between the contour holes and the nearest production holes are kept high to facilitate complete movement of material in production holes before the contour holes detonate so that the gas expansion in contour holes occurs towards the opening. Sometimes, holes are drilled in between two charged blast holes and are kept uncharged. These are called dummy holes (Figure 1). The stress concentration at the farthest and the nearest points of the dummy holes become high to initiate cracks from the dummy holes extending towards the charged holes. The fracture is, thus, controlled along the desired contour.

In some cases, slashing or trimming techniques are used where the central core of excavation area is removed first to reduce the stress and then post-splitting is adopted to remove the remaining rock mass along the desired contour. The technique is generally referred to as 'slashing' or 'trimming' [Calder and Bauer (1983), Figure 2].

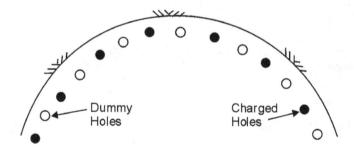

Fig. 1. Smooth blasting pattern with dummy holes

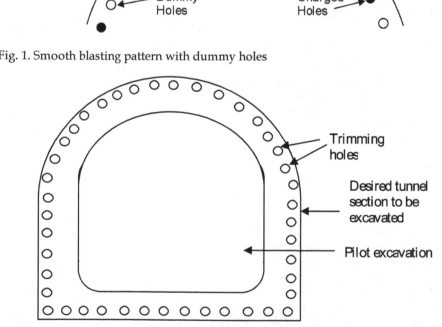

Fig. 2. Cushion blast holes for trimming of a tunnel after pilot excavation

Line drilling is adopted as an alternative technique where a number of uncharged holes are drilled along the contour with a spacing of 2-4 times the hole diameter (Du Pont, 1977). The distance of the row of empty holes from the final row of charged holes is kept as 0.5-0.75 times the normal burden. The empty holes are joined during the main blasting round and a separation is created along the contour.

According to Holmberg and Persson (1978), the spacing of pre-split holes should be 8-12 times the blast hole diameter. The following design parameters for contour hole spacing, burden to spacing ratio of contour holes and linear charge concentration in smooth blasting are suggested by Holmberg (1982) :

$$S_{dc} = 16 \times d_b, \text{ m} \tag{5}$$

$$m_{dc} = 1.25 \tag{6}$$

$$q_{lcc} = 90 \times (d_b)^2, \text{ kg/m} \tag{7}$$

where

S_{dc} = spacing of contour holes while drilling, m,
m_{dc} = burden to spacing ratio of contour holes while drilling,
q_{lcc} = linear charge concentration in the contour holes, kg/m, and
d_b = diameter of blast holes, m.

Controlled blast design details recommended by Olofsson (1988) are presented in Table 3.

Type of Blasting	Blast Hole Diameter (mm)	Spacing of Blast Holes (m)	Burden (m)	Linear Charge Concentration (kg/m)
Smooth blasting	25-32	0.25-0.35	0.3-0.5	0.11
	25-48	0.5-0.7	0.7-0.9	0.23
	51-64	0.8-0.9	1.0-1.2	0.42-0.45
Pre-splitting	38-44	0.3-0.45	-	0.12-0.37

Table 3. Recommended blast design for contour blasting (Olofsson, 1988)

7. Special tunnel blasting techniques

Some special blasting techniques were practised in tunnels and underground coal mine to attain greater advance and better safety in some critical working sites under the recommendations and supervision of Central Mining Research Institute, Regional Centre, Nagpur. Those cases are discussed in brief in the following paragraphs.

7.1 Long hole raise driving by blasting

A 123 m deep pilot shaft was excavated in 95 days time using Long Hole Raise Blasting (LHRB) method for faster and safer shaft sinking in the surge shaft, passing through various kinds of basaltic formations, in Ghatghar Hydro-electric Project of Maharashtra. The blast hole charging pattern is shown in Figure 3. Application of this techniques resulted in saving of 75% time 60% cost of excavation in comparison to the conventional shaft sinking method.

A ventilation shaft of 40m depth was also excavated by using the same technique at diversion tunnel of Latur-Osmanabad Railway tunneling project of Central Railways in 20 days. This techniques yielded in saving of time by 80% and cost of excavation by 60% in comparison to the conventional shaft sinking method, which mainly suffer from weather effects, confined working space and low cycle time.

Similarly, a pilot surge shaft of 3.0m diameter130m depth was excavated by long hole raise driving technique at a lift irrigation scheme of Koilsagar project. This swift and cost effective shaft excavation technique was completed in just 60 days with cost savings of 70% and time saving by 95% in contrast to conventional shaft sinking method. The profile of excavated pilot surge shaft at Koilsagar project is shown in Figure 4.

7.2 Lake tap blasting

The lake taping of fist of its kind with indigenous technology was carried out in India by CMRI (now CIMFR) at granitic rock mass in South India. The Andhra Pradesh Power Generation Corporation (APGENCO), India, executed a lift irrigation scheme (SLBC) for the Government of Andhra Pradesh to install 4 Nos. of 4 x 25000 hp pumps to lift 2400 cusecs of water from the Nagarjuna Sagar reservoir for irrigation purpose. A 4 m thick rock plug, designed by CMRI, was left for lake tapping at the end of project. The area of cross section of the tunnel was 40 m². Considering proximity of the nearby structures a controlled blast strategy in phased manner was evolved prior to final plug blasting. Vibration and damage characteristics were ascertained to finalise the blast design of the final plug.

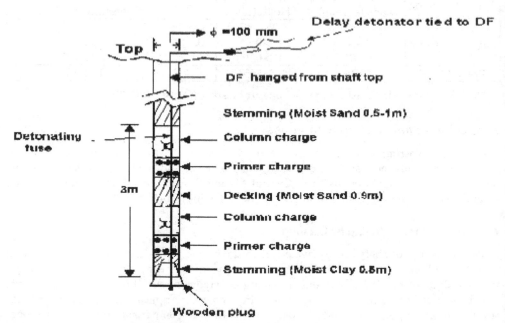

Fig. 3. Charging pattern in raise blasting

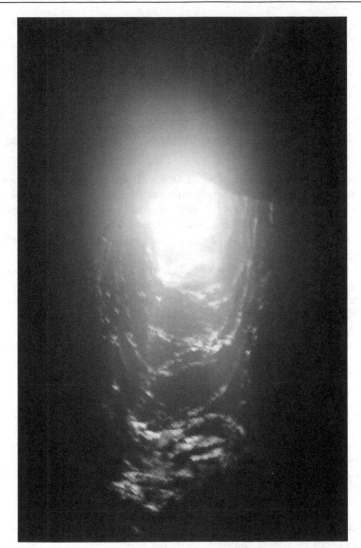

Fig. 4. Profile of pilot surge shaft excavated by long hole raise driving at Koilsagar

Based on the blast performance of the trail rock plug final plug blast design was made with the following salient features:

- Specific charge was increased from 1.25 kg/m³ to 1.33 kg/m³ to improve throw and fragmentation.
- Only gelatine explosive was recommended considering the water inflow from the blast holes.
- Dummy holes were made above the crown holes, at a distance of 0.3 m, to minimise rock mass damage.
- A borehole from the top was used to convey initiation to the blast holes.

The final plug-blasting pattern is shown in Figure 5. This novel technology being an indigenous one could save Crores of national exchequers.

7.3 Cautious blasting

By adopting an extremely cautious approach, all 10 reinforced concrete plugs, each of 125 m^3 volume, in 5 units were removed by controlled blasting without causing any damage to the surrounding periphery and pier nose in Srisailam left bank project of the APPGENCO while the power house was in running condition. The controlled blasting pattern is described below:

i.   Line drilling holes of 1.5m depth were drilled with spacing of 0.15 m between the holes on the pier nose side and at 0.20 m inside the periphery.
ii.   The periphery holes were pre-split with air-decking. The half cast factor of the periphery blasting was around 95%, which indicates low damage level. The pre-split blasting connections and the post-blast wall with half cast holes are shown in Figures. 5(a) and 5(b).
iii. A cut was created at the heading and it was widened and deepened to make a pilot hole in the plug along its axis.
iv.   The balance concrete mass of the heading was slashed with less charge against the void.
v.   The bottom was blasted with benching method.
vi.   Mucking was done by mechanical and manual means.
vii. Continuous blast vibration monitoring was carried out during the blasts at near, intermediate and far field.
viii. Analysis of vibration data was done for subsequent blasting and to develop general predictor equation.

Pre and post blast ultrasonic measurements were taken at the exposed areas of the pier nose walls to know the change in physical property the reinforced mass due to blasting. The compressional wave velocities (P-wave) were measured by Roop telesonic ultrasonix instrument 'Ultrasonix 4600' which is shown in Figure 6. The average P-wave velocity was 2075 m/s and 2100m/s before and after blasting respectively. The values indicate that there has been no blast-induced damage to the structure under consideration.

The cautious blasting was also applied at Koldam Hydroelectric Power Project (KHEPP) to reduce overbreak and to get a smoother tunnel wall profile. The rock mass encountered in all the tunnels of KHEPP was Dolomite, which was very heterogeneous, highly weathered, metamorphosed, compact, foliated, sheared and crushed due to the effect of Chamiatar Khad fault striking N1700 E and 450 W. Joints are open, closely spaced, intersecting, which are having clay fillings due to mechanical and chemical weathering of the rocks. One main joint with angle of N 750 E/800W is running parallel to the axis of the tunnels which is very unfavourable. At some places huge wedges were formed due to the intersection of the joints, which caused excessive overbreaks in the tunnels. The Q values of most of the rock mass of tunnels range from 0.12 to 0.21, which indicates that the rock was very poor. Core samples were collected from both the monitoring locations by underground coring machine. Engineering properties like Rock Quality Designation (RQD) compressive strength, tensile strength, density and compressional wave velocity (Vp) were determined from the core samples.

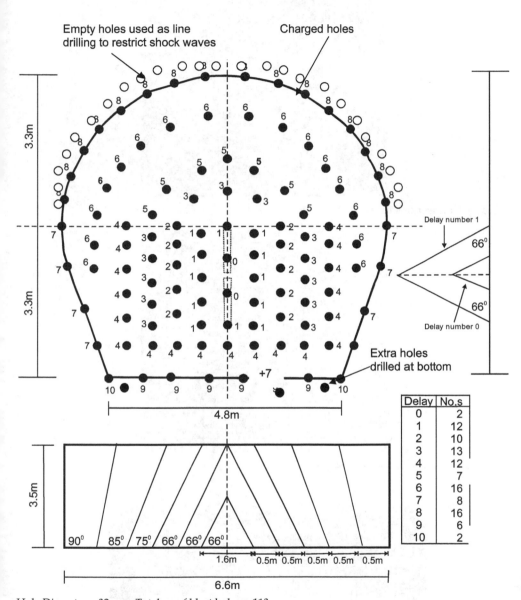

Fig. 5. Lake tap blast design

Hole Diameter = 32 mm; Total no.of blast holes = 113,
Length of blast holes = 3.5 to 4.0 m,
Specific Charge = 1.33 kg/m3

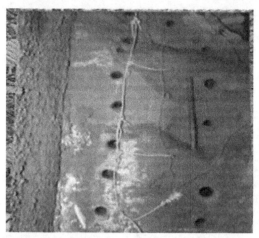

Fig. 5(a). Connections for pre-split blasting

Fig. 5(b). Pier nose wall after pre-split blasting

Fig. 6. Compressional wave velocity (P-wave) measuring device 'Ultrasonix 4600'

In-situ compressive strengths were also determined by using Schmidt hammer rebound testing. The average RQD values of Dolamite rock mass ranging from of 40-60%. Water absorption properties measured at the test site was 1.2% at both the sides. The improved blast performance of smooth blasting in the form of smooth profile is shown in Figure 7. The results were consistent for 12 trial blasts at the Dolomite tunnel. The controlled blasting restricted the overbreak to only 3%, which was 27% with the conventional tunnel blasting. The average half cast factor was calculated as 85%.

Fig. 7. Improved blast performance of smooth blasting in the form of smooth profile at KHEPP

7.4 In-hole delay blasting

Following the trend of opencast blasting, in hole delay blasting technique using delay electric detonators were used in some mines and tunnels to improve the pull per blast and reduce the ground vibration. As the confinement in the cut holes are maximum and the blast performance in tunnels depend mainly on the development of the cut portion, the in-hole delay were used in the cut holes only. The salient features of the in-hole delay pattern are:

1. The collar portion of the hole was blasted prior to the bottom. Thus, the confinement at the hole bottom was less during firing.
2. Mid-column decking between the two charges in a hole was kept at least 0.6 m to avoid sympathetic detonation. This decking provided confinement for the bottom charge.

The charging pattern is explained in Figure 8.

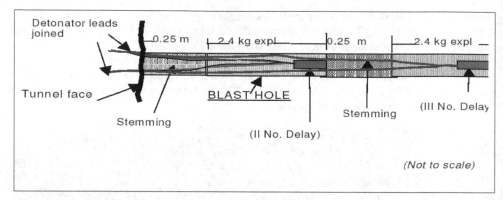

Fig. 8. Charging pattern of cut holes with in-hole delay

This technique was successfully applied at basaltic rock mass of Central railway tunnels and gneiss rock mass of Lohari Nag Pala Hydel power project.

The advantages of the in-hole delay cut blasting includes:

1. The average face pull improve by nearly 30-50%. The specific charge also reduces proportionately.
2. The blast vibration intensity reduces by 20 to 25% as the cut hole charge is distributed in two delays. This is going to reduce the overbreak proportionately.

7.5 Bottom hole decking technique

The mining industry is striving to enhance the productivity by improving fragmentation to reduce the system cost. In order to achieve this objective, development of new techniques and their application is essential. The authors at CIMFR, experimented a blasting technique called 'bottom hole decking technique' to achieve the objective of blasting productivity improvement of the mining industry. The technique consists of air decking at the bottom of the blasthole in dry holes by means of a wooden spacer or a closed PVC pipe. Although, practice of air decking is not new thing in blastholes, the concept of inserting bottom hole decking below the explosive column is relatively new. Explosives provide a very concentrated source of energy, which is often well in excess of that required to adequately fragment the surrounding rock material. Blast design, environmental requirements and production requirement limits the degree to which the explosive energy distribution within the blasthole can be significantly altered using variable loading techniques. Use of air-decks provide an increased flexibility in alteration and distribution of explosive charge in blast holes.

The bottom hole air-decking was developed to avoid the general disadvantages of middle air decking and to simplify the complex charging procedure, associated with it. The design aspects of the technique are explained in the following sections. The bottom hole decking consists of air decking at the bottom of the hole in dry holes by means of a spacer or a closed PVC pipe, covered at the upper end. The fume characteristics of the spacer are to be tested before applying in underground coal mine. If blast holes are wet, water decking will be created at the bottom by means of a spacer with a weight attached to it for sinking to the bottom. The diameter of the spacer should be preferably one third of the blasthole diameter for easy lowering and not allowing the charge to go to bottom side while loading. The reported values of air-deck length was taken as basis for optimum bottom deck length which was about 10% of the hole depth (Mead et al, 1993). The hole contains explosive and stemming column as in conventional loading but with a spacer at the bottom. The principle of bottom hole air decking in achieving optimum explosive energy interaction on rock mass is given below:

- Reduced shock energy around the blast hole due to cushioning effect of air decking, which otherwise would result in crushing
- Explosive energy-rock interaction is more at the bottom due to relative relief zone existing at that zone.
- Effective toe breakage is due to striking and reflection of shock waves at the bottom face of hole

The procedure and sequence of blast hole loading and initiation for the bottom hole decking are given below:

- Inserting the spacer in to the hole bottom by stemming rod.
- Loading the primer explosive cartridge attached by delay detonator charging the column charge conventionally
- Stemming of the hole by proper stemming material, preferably by sand mixed clay

The advantages of the bottom air decking technique in comparison to the conventional middle air decking are given below:

i. The highly confined toe is free of explosive charge but exposed to high concentration shock energy, resulting in good toe breakage and low vibration intensity.
ii. The reduced overall peak shock reduces the back break and damage.

Blast hole charge design for production blasts with bottom air-decking is Figure 9.

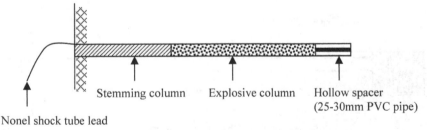

Stemming column   Explosive column   Hollow spacer
                                             (25-30mm PVC pipe)

Nonel shock tube lead

Fig. 9. Blast hole charge design for production blasts with bottom air-decking

The bottom air decking also resulted in the overall progress/pull per round of 36% with 1.5 deep rounds and 22% with 1.8 m deep rounds even with the powder factor improvement (ton/kg) upto 70%. The increase of detonator factor was very predominant in case of tests with bottom decking in comparison to tests with bottom decking technique. The technique was also resulted in reduction of ground vibrations by 20-26%. The laboratory and field experimental results prove that the bottom-hole air decking is an effective technique for improving the opencast blasting productivity as well as safety.

8. Sand stemming device for horizontal blast holes

The device of sand stemming for horizontal blast holes constitutes an assembly of a plastic pipe with proper cut and slits, a wooden block with pulley arrangement for resisting sand and an anti-static (non metallic) rope to pull out the plastic pipe from blast hole. The device essentially consists of a plastic pipe tied with an anti-static rope which is passed through a wooden resisting block to which a pulley is attached. The main objective of the device is for efficient use of sand as stemming material in horizontal blast holes. Another objective of the present device is to provide an effective and economic and fast stemming method which can find a mass application in underground blasting. The device essentially consists of a plastic pipe tied with a non metallic rope which is passed through a wooden resisting block to which a pulley is attached, an assembly of a plastic pipe cut and slit properly and a rope passed through a wooden block which can insert the sand in to blast hole and resist the sand to come out while pipe is pulled out of the blast hole. The position of stemming device while inserting the sand with plastic pipe and the position of removal of the plastic pipes are shown in the Figure 10. Actual application of the device in the field is shown in the Figure 11.

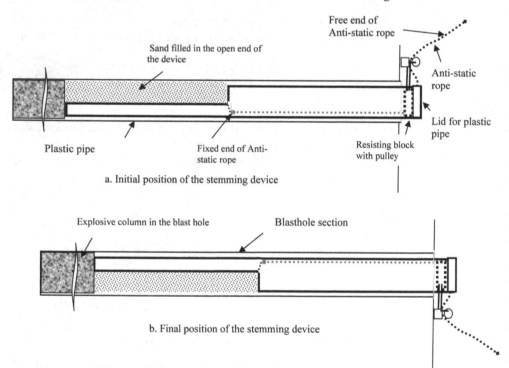

Fig. 10. Position of stemming device for loading and unloading in the blasthole.

Application of this tool in place of conventional stemming resulted in pull improvement of 5-10% in dolomite tunnels and 8-12% in gneiss tunnel. The improved blast performance was recorded consistently for 20 trial blasts at the gneiss tunnels and 25 trial blasts at the dolomite tunnels.

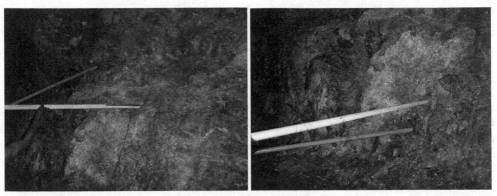

Fig. 11. Loading and unloading of sand into the stemming column of a blasthole at KHEPP

9. Computer aided blast design

Some of the software developed for blast design and optimisation are reported in Table 4. Few blasting software on tunnel blasting are commercially available and the details can be obtained through web search.

Name of Software	Purpose	Reference
OPTES	Blast optimisation in tunnels	Vierra (1984)
VOLADOR	Estimation of blast results, blast efficiency and cost analyses in tunnels	Rusilo et al. (1994)
TUNNEL BLAST	Blast design in tunnels	Chakraborty et al. (1998)
CAD	Optimum design of ring hole blasting	Myers et al. (1990)
FLAC and UDEC	Blasting effects on the near field rock mass	Pusch et al. (1993)
ABAQUS V 5.4	Mechanics of crater blasting and the effects of air decking and decoupling	Liu and Katsabanis (1996)
ALEGRA	Air-decking blasting	Jensen and Preece (1999)
PFC-2D/3D	Crack and heaving simulation	Itasca Consulting Group Inc. (2002)
Neural networking	Model free computing	Leu S. S. et al. (1998)

Table 4. Various routines for computer aided tunnel blast design

9.1 TUNNEL BLAST$^{2.0}$ software

Based on the past experience and extensive field investigations over a variety of underground structures of varying lithologies, CMRI Nagpur Centre devised a software "TUNNELBLAST" for generating blast design for Tunnels and underground workings. This software is a handy intelligent tool for the site engineers to optimise the blasting process and improve productivity without spending their valuable time on scrutinising variety of documents, books and literature available. The software is simple to operate and user friendly. The input and output parameters of the software are as under:

Input parameters:

1. Rock properties (density, compressive strength and joint spacing),
2. Tunnel (shape, width and height),
3. drilling (diameter and length of blast hole), and
4. Explosive properties (weight strength, weight and length of cartridge).

Output parameters:

1. Size of the tunnel,
2. Probable deviation of blast holes,
3. Optimum depth of round,
4. Look out angle of peripheral holes,
5. Burden, spacing and charge of holes in cut area, floor periphery and in the middle of the tunnel section,
6. Front and sectional views of the blast pattern

9.2 Field application of TUNNEL BLAST software at gneiss rock mass

The TUNNEL BLAST software was applied to design the parallel cut blast pattern at Lohari-Nag Pala Hydroelectric Power Project (LNPHPP). The LNPHPP falls in the Uttrakhand Himalayas and is located on the River Bhagirathii upstream of Uttarkashi district. The main rock type of powerhouse complex is schistose gneiss and augen gneiss with abundance of mica and geotechnically the rock mass is negotiating in "Fair Category" and it's having three prominent joint sets. The Rock Mass Quality (Q) was varying from 1-10. The main two joint sets intersecting at right angle which makes wedge continuously. Some weak zone/clay filling, altered rock, sheared rock mass and excessive flow of water at places makes the rock poor. In maximum area it is found that the regional trend of foliation is perpendicular to the tunnel alignment, another joint which is intersecting the foliation at right angle and creates wedge on roof. The strike of the foliation is going through along the tunnel alignment which is geologically not favourable because of probabilities of plane failure and wedge failure in presence of heavy joint planes.

The input geological parameters required for the blast design software are as follows:

Rock type	Metabasic (Amphibolite) & quartz vein
Joint sets	Three + Random ($045^0/35^0$, $210^0/45^0$, $130^0/80^0$)
Critical joint	$045^0/35^0$, $210^0/45^0$
Water Condition	Dry
Weathering	Highly Weathered/Fractured
Filling	Clay seam, width 10-20cm

Boundary conditions:

- Rock Type: Metabasic (Amphibolite) & quartz vein
- Av. rock density: 2.6 t/m³
- Type of explosives: Emulsion
- Blast hole diameter: 45 mm
- Explosives diameter: 40 mm & 32 mm
- Explosives strength: 80% (60% may also be required in the periphery holes
- and hence provisions may be made)
- Length of blast hole: 2 m
- Delay: Long delay (NONEL)

After feeding the input information the software process the entire data and gives the blast hole geometry and charge pattern for cut holes and other holes separately. The utput information given by TUNNELBLAST software is given in Figure 12, Figure 13 and Table 5 and Table 6.

The blast design generated by TUNNEL BLAST software was applied at intermediate adit and the blast results were satisfactory in terms of pull per round and overbreak control. The trial blast results with felid application of TUNNEL BLAST software are given in Table 7. The blast results indicate the efficacy of the TUNNEL BLAST software, as a preliminary tool for tunnel blast design for various geological conditions. The fine tuning of this design can be done for further improvements in the progress and yield of tunnel blasting.

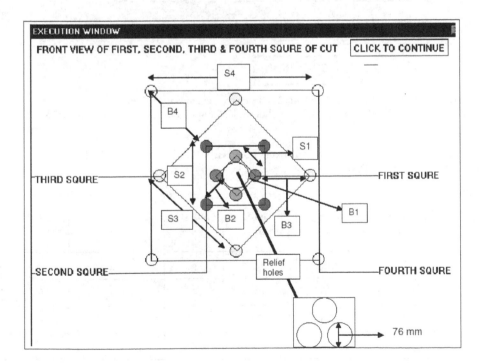

Fig. 12. Blast design output from TUNNELBLAST for cut holes of intermediate adit

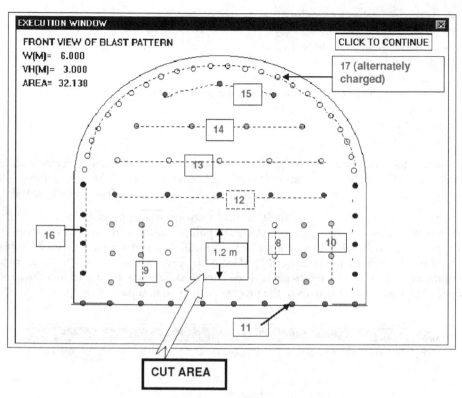

Nos. in the boxes denote the delay numbers; Total Charge per round = 97.7 kg
Total no. of holes= 3-Relief holes + 69-Charged holes+ 12-Dummy holes; Powder factor = 1.52 kg/m³

Fig. 13. Controlled blast design output from TUNNEL BLAST for rest of the holes at intermediate adit of LNPHPP

Short Delay No. (25 ms delay)	Name of square	Burden, m	Spacing, m	No. of holes	Charge/hole, kg	Total charge, kg
1	First	0.15	0.2	4	1.2	4.8
2/3	Second	0.20	0.4	4	2.4	9.6
4/5	Third	0.35	0.75	4	2.4	9.6
6/7	Fourth	0.45	1.2	4	2.4	9.6

Table 5. Blast pattern and charge configuration of the cut holes

Description of holes	Delay No.	No. of holes	Burden	Spacing	Charge per hole (kg)	Total Charge/ delay (kg)
Easer holes	8	6	0.6	0.75	0.95	5.7
Support holes	9	6	0.6	0.75	0.95	5.7
Support holes-II	10	6	0.6	0.75	0.95	5.7
Bottom Holes	11	10	0.4	0.70	1.8	18
Crown Holes-I	12	5	0.8	1.2	1.6	8
Crown Holes-II	13	5	0.8	1.2	1.6	8
Crown Holes-III	14	4	0.8	1.2	1.6	6.4
Crown Periphery holes	15	3	0.6	0.3	0.6	1.8
Side Periphery holes	16	8	0.6	0.3	0.6	4.8

Table 6. Design and charging details of blast holes, other than cut holes

S No.	Location	Hole diameter, mm	Depth of holes, m	No. Of holes	Charge per round, kg	Specific charge, kg/m^3	Pull/round, m
1	Downside, TRT	40	3.5	91	217	2.1	1.98
2	Upsideside, TRT	40	3.5	89	225	1.85	3.1
3	Upsideside, TRT	40	3.5	89	250	1.98	3..0
3	Downside, TRT	40	3.5	91	220	2.0	1.95

Table 7. Trial blast results with felid application of TUNNEL BLAST software

10. Conclusions

The reviews on the developments in rock mass damage and contour blasting brings an important information on field application of controlled blasting and damage assessment and control The contributions of CIMFR on special tunnel blasting techniques resulted in improvement of both productivity and safety. The following conclusions can be drawn based on the various topics discussed in the paper:

i.   Application of this techniques resulted in saving the time of 75-80% and cost of 60%-95% in comparison to the conventional shaft sinking method at three different projects

ii.   Lake Tap Blasting of a 4 m thick 40 m2 cross sectional area was carried out as of fist of its kind with indigenous technology in India by CMRI (now CIMFR) at granitic rock mass in Andhra Pradesh Power Generation Corporation (APGENCO), which could save Crores of national exchequer.

iii. Ultra cautious blasting techniques were adopted as an extremely cautious approach, for removal of 10 reinforced concrete plugs, each of 125 m3 volume, without causing any damage to the surrounding periphery and pier nose in Srisailam left bank project of the APPGENCO while the power house was in running condition.

iv.   Successful application of in-hole delay cut blasting method at basaltic rock mass and gneiss rock mass improved average face pull improve by nearly 30-50%. Blast vibration intensity reduces by 20 to 25% which resulted in reduction of the overbreak proportionately.

v.   Bottom hole decking technique resulted in the overall progress/pull per round of 36% with 1.5 deep rounds and 22% with 1.8 m deep rounds even with the powder factor improvement (ton/kg) upto 70%.

vi.   Application of sand stemming device for horizontal blast holes in place of conventional stemming resulted in pull improvement of 5-10% in dolomite tunnels and 8-12% in gneiss tunnel.

11. References

Bieniawski, Z. T., (1993). Classification of rock masses for engineering: the RMR system and future trends. In J.A. Hudson (Ed.), Comprehensive rock engineering: principles, practice, and projects, Oxford: Pergamon Press. V.4, pp. 553-573

Calder, P. N. and Bauer, A. (1983). Presplit blast design for open pit and underground mines, 5th International Cong. on Rock Mechanics, Melbourne, Vol. 2, pp. E185- E190.

Chakraborty, A. K. (2002). Development of predictive models for optimum blast design in mine roadways and tunnels under various rock mass conditions, Ph.D. Thesis, Indian School of Mines, Dhanbad, India, 298 p.

Chakraborty, A. K., Jethwa, J. L. and Dhar, B. B. (1996b). Predicting powder factor in mixed-face condition: development of a correlation based on investigations in a tunnel through basaltic flows, Engineering Geology, Elsevier Science B.V., Netherlands, No. 47, pp. 31-41.

Chakraborty, A. K., Murthy, VMSR and Jethwa, J.L. (1996a). Blasting problems in underground construction through Deccan Trap formation: Some experiences at Koyna Hydro-electric Project, Stage IV , Tunnelling & Underground Space Technology, Elsevier Science Ltd., Great Britain, Vol. 11, No. 3, pp. 311-324.

Chakraborty, A. K., Murthy, VMSR, Jhanwar, J. C., Raina, A. K., Ramulu, M. and Jethwa J. L. (1998). Final report on Development of Rock Mass Classification for Computer Aided Tunnel Blast Design, Grant-in-Aid Project funded by Ministry of Water Resources, Govt. of India, 136 pp.

du Pont, E.I., (1977). Blasters hand book, 175th Anniversary edition, E.I. du Pont de Nemours, Inc., Wilmington, Delaware. Pp.526-541.

Duvall, W. I. and Atchion, T. C. (1957). Rock breakage by explosives, USBM, RI 5356, Explosives Reference Database on CD-Rom, International Society of Explosives Engineers, Ohio, USA, 1997.

Gupta, R. N., Singh, R. B., Adhikari, G. R. and Singh, B. (1988). Controlled Blasting for Underground Excavation, International Symposium on Underground Engineering, 14-17 Apr., New Delhi, India, pp. 449-460.

Holmberg, R., and Persson, P. A. (1978). The Swedish approach to contour blasting, Proc. of Annual Conference on Explosives and Blasting Research, Explosives Reference Database on CD-Rom, International Society of Explosives Engineers, Ohio, USA, 1997.

Itasca Consulting Group Inc. (2202). Partcile Flow Code in 2 Diemnsions – Theory & Background, Mnneapolis, USA. p. 1.1-1.28.

Jensen, R. P. and Preece, D. S. (1999). Modelling explosive/rock interaction during pre-splitting using ALE computational method, *Proc. 6th. International Symposium for Rock Fragmentation by Blasting*, The South African Inst. of Mining and Metall., Johannesburg, Aug. 8-12, pp. 199-202.

Johansen, J. (1998). Modern trends in tunnelling and blast design, IDL Industries Ltd., Hyderabad, India, pp. 34-41.

Langefors, U. and Kihlstrom, B. (1973). *The Modern Technique of Rock Blasting*, John Willey & Sons, pp. 188-257, 299-301.

Leu S. S., Lin S. -F., Chen C. -K. And Wang S. -W. (1998). Analysis of powder factors for tunnel blasting using neural networks, The Int. Journal for Blasting and fragmentation, Balkema A. A. , Netherlands, Vol. 2., No. 4, pp. 433-448.

Liu, L. and Katsabanis, P. D. (1996). Numerical modelling of the effects of air-decking/decoupling in production and controlled blasting, *Proc. 5th International Symposium on Rock Fragmentation by Blasting, FRAGBLAST-5*, Montreal, Quebec, 25-29.,pp. 319-330.

Livingston, C. W. (1956). Fundamentals of rock failure, *Quarterly of the Colorado School of Mines*, Vol. 51, No. 3, Jul..

Lopez Jimeno, C., Lopez Jimeno, E., Carcedo, F. J. A. and De Ramiro, Y. V. (1995). *Drilling and Blasting of Rocks*, Balkema A. A., Rotterdam, pp. 200-204 and 259-260.

McKenzie, C. J. (1994). *Blasting for Engineers*, Blastronics Pty. Ltd., Brisbane, Australia.

Olofsson, S., O. (1988). Applied Explosives Technology for Construction and Mining, Applex, Arla, Sweden, 303 pp.

Ouchterlony, F., Nyberg , Sjoberg, C., Johansson, S-E. (1991). Damage zone assessment by vibration measurements, *Aspo hard rock laboratory*, Progress Report, , No. 3, pp. 25-91.

Pokrovsky, N. M. (1980). *Driving Horizontal Workings and Tunnels*, Mir Publishers, Moscow, pp. 38-41.

Pusch, R. and Stanfors, R. (1992). The zone of disturbance around blasted tunnels at depth, *International Journal Rock Mech. Mining Sci. & Geomech. Abstr.*, Vol. 29, No. 5, pp. 447-456.

Pusch, R., Hokmark, H. and Borgesson, L. (1993). Characterisation of structure and stress state of nearfield rock with respect to the influence of blasting, *Proc. 4th International Symp. on Rock Fragmentation by Blasting, FRAGBLAST-4*, Vienna, Austria, pp.175-181.

Ramulu, M., (2009), Rock mass damage due to repeated blast vibrations in underground excavations, IISc Dept.of CE PhD Thesis- G23635, 624.15132 P09, IISc Press.

Ramulu, M., Chakraborty A. K. and Sitharam T.G., (2009), Damage assessment of basaltic rock mass due to repeated blasting in a railway tunnelling project – a case study, Tunnelling and Underground Space Technology, Vol.24, pp. 208–221.

Ramulu, M., Sitharam, T.G., (2011), Blast induced rock mass damage In underground excavations -Applications to civil engineering projects, LAMBERT Academic Publishing GmbH& Co. KG, 66123 Saarbrücken, Germany, ISBN (978-3-8433-9318-8)

Rusilo, L. C., Sansone, E. C., Hennies, W. T. and Ayres Da Silva, L. A.(1994). Design and optimization of tunnel blasting operations, *Proc. 3rd International Symposium on Mine Planning and Equipment Selection*, Istanbul, Turkey, 18-20 Oct., pp. 651-655.

Singh, S. P. (1995). Mechanism of cut blasting, Trans. Inst. Mining and Metall., Section A, Mining Industry, Vol. 104, Sept-Dec.,*The Inst. of Mining and Metall.*, U.K., pp. A134-A138

Wilson, W. H. (1987). An experimental and theoretical analysis of stress wave and gas pressure effects in bench blasting, *Ph.D. Thesis*, University of Maryland.

Yu, T. R. and Vongpaisal. S. (1996). New blast damage criteria for underground blasting, *CIM Bulletin*, No. 998, Vol. 89, pp. 139-145.

## 8

Susceptibility of the GSM-R Transmissions to the Railway Electromagnetic Environment

Stephen Dudoyer[1], Virginie Deniau[1],
Nedim Ben Slimen[2] and Ricardo Adriano[3]
[1]*Univ Lille Nord de France, IFSTTAR,*
[2]*Assystem,*
[3]*Federal University of Minas Gerais,*
[1,2]*France*
[3]*Brazil*

1. Introduction

The Electromagnetic radiations are non-ionising radiations; they cannot involve the ionisation of atoms or molecules. Nevertheless, they can cause various adverse effects. From a biological point of view, they provoke heating due to the occurring of induced current in the body. But, this issue will not be considered in this chapter. From a technological point of view, they can cause malfunctions, permanent damages for electronic devices or telecommunication systems. In this chapter we will focus on their impact on a telecommunication system dedicated to the European railway and potential consequences on the management of the railway network.

Today, the European railway network is undergoing significant changes, which aim at deploying a unique management system in Europe which will replace local systems. This unique management system called ERTMS (European Railway Traffic Management System), involves the deployment of a telecommunication network dedicated to railway management, the GSM-Railway network, in order to harmonize in Europe the system of communication between the trains and the infrastructures. This harmonization is intended to clear the technological boundaries between railway networks of European countries and thus to remove border for trains. GSM-R is a key element in the management system as it provides the vocal exchanges, but also the transmission of signalling data. However, as all the telecommunication systems, the GSM-R can be vulnerable to the Electromagnetic (EM) interferences and the railway environment is particularly rich in EM interferences. This chapter will then focus on this issue.

After a general background about the electromagnetic interferences and the management of the European railway network, we present the standards and approaches applied in the railway domain to control the Electromagnetic compatibility (EMC). The GSM-R and the EM disturbances which can affect it are then detailed. Finally, a methodology for testing the vulnerability of the GSM-R transmissions and the test results are presented and analysed.

2. General notions

Understanding the electromagnetic emission from the railway environment is important to prevent and control electromagnetic interference. Currently, trains are more and more often equipped with potentially sensitive systems from an electromagnetic compatibility point of view. Consequently, railway systems have to be sufficiently robust to guarantee the safety of the railway transportation. In this section the fundamental concepts related to EMC are briefly introduced. For this purpose, the following definitions given in (IEC 60050, 1990), International Electrotechnical Vocabulary (IEV), chapter 161, apply:

Electromagnetic environment: The totality of electromagnetic phenomena existing at a given location.

Immunity (to a disturbance): The ability of a device, equipment or system to perform without degradation in the presence of electromagnetic disturbance.

(Electromagnetic) Susceptibility: The inability of a device, equipment or system to perform without degradation in the presence of an electromagnetic disturbance.

Immunity level: The maximum level of a given electromagnetic disturbance incident on a particular device, equipment or system for which it remains capable of operating at a required degree of performance.

2.1 Electromagnetic disturbances and electromagnetic compatibility

A system is electromagnetically compatible with its environment if it is able to operate satisfactorily in its electromagnetic environment without introducing intolerable electromagnetic disturbances to anything in that environment. Typically, an EMC problem can be decomposed into three main parts. A source that intentionally or unintentionally produce the emission, a coupling path that transfers the emission energy to a receptor and the receptor that can be susceptible if the received energy exceeds its immunity level.

During their operation, electrical or electronic systems generally produce radiated or conducted signals, which can lead to equipment malfunctions neighbours. The "electromagnetic disturbance" term then assigns these signals that can be voltages, currents or electromagnetic fields. In general, the higher the frequency of the electromagnetic disturbance is, the more efficient the coupling path. It is important to keep in mind that the source and receiver can be classified as intend or unintended. For instance, the GSM-R system intentionally transmits and receives electromagnetic fields in some frequencies between 876 MHz and 925 MHz. Consequently, the equipment near the GSM-R antennas must be designed to operate properly under the influence of the GSM-R signals. On the other hand, the GSM-R antenna will collect all the signals generated by the railway environment at these frequencies. Depending on the coverage of the GSM-R system and the levels of the electromagnetic disturbances, the communication between rolling stocks can be affected or even interrupted.

2.2 Electromagnetic coupling

Electromagnetic disturbances produced by the emitter can be coupled to the receptor by either radiated or conductive paths. The coupling mechanism can be classified into Conductive coupling, Magnetic coupling, Electric field coupling and Electromagnetic field coupling.

Conductive coupling can be viewed as a common impedance coupling. Conductive coupling occurs when the source and the receptor circuits are physically connected with a conductor and share a common-impedance path. Magnetic coupling occurs when two objects exchange energy through their varying or oscillating magnetic fields. It can be represented by a mutual inductance between the source and the receptor. Electric field coupling (or capacitive coupling) is caused by a voltage difference between conductors. It dominates in high-impedance circuits and can be represented by mutual capacitance. Finally, Electromagnetic coupling is a combination of both electric and magnetic fields. It is the most common coupling mechanism observed. It occurs when source and receptor are separated by a large distance, (typically more than a wavelength). In this case, source and receptor act as radio antennas. The electromagnetic field radiated by the source propagates across coupling path and is picked up by the receptor.

2.3 Electromagnetic radiation, emission and immunity

Any device which drives an electric current is likely to radiate an electromagnetic field. This electromagnetic field may act in two ways on electronic and telecommunication equipment. It can either be picked up by cables of other systems (or lines of electronic circuits) on which unwanted electrical signals appear and can cause malfunctions or it can also interfere with the telecommunication signals when they reach the receivers causing losses of information. In the first situation, the interference occurs because the dimensions of the conductors in the electronic equipment are comparable with the wavelength of the electromagnetic disturbance. In this case conducting elements can act as receiving antennas.

There are numerous sources of unintentional electromagnetic radiation such as lighting, relays, electric motors and digital systems. The number of emitters is increasing rapidly. Some of these emitters employ very high power levels; others such as digital systems are using faster digital electronics and are becoming more efficient radiators of unintentional electromagnetic energy. Consequently, EMC has become a particularly important topic. In order to ensure that EMC will be not a problem, many EMC standards are used. These are often supported by EMC legislation to ensure that existing equipment conforms to the required standards. EMC standards specify a limited number of essential emission and immunity tests, as well as minimum test levels. The aim is to ensure adequate compatibility. Section 4 summarizes the major standards concerning the electromagnetic emissions in railway environment while section 5 addresses the immunity problems.

3. Management and signalling of european railway network

The management of a railway network is generally performed thanks to several key components, notably a ground-train radio which allows the vocal exchanges, a lateral signalling system including lights and traffic signs and a localization system of the trains which can also control the speed of the trains. However, these different components are not necessary ensured by similar technologies in all the European countries. This situation inhibits the carrying out of a real European railway network which would allow the different railway operators to offer their services anywhere in Europe. Today, trains crossing borders are necessarily equipped with various national systems and at the borders the trains have to change their system to be in accordance with the cross border country. This increases the costs of equipment and maintenance of the trains, the operating costs and extends the travel time.

The ERTMS (European Railway Traffic Management System) standard was then thought out in order to remove these obstacles and to optimize the use of the European railway network and to improve the reactivity, adaptability and affordability of the European railway. ERTMS would allow the interoperability of trains on the European territory (Jarašūnienė, 2005). This standard is generally presented as composed of two main components, which are the European Train Control System called ETCS, a standard for in-cab train control, and the GSM-R (Global System for Mobile communications-Railway) system, an international wireless communications standard dedicated to railway applications.

ETCS can allow automatically controlling the speed of the train if necessary. ETCS is composed of trackside and on-board modules. The trackside module transmits information to the train which enables the on-board computer, called Eurocab, to calculate the maximum permitted speed.

Nevertheless, the implementation of ETCS requires major adjustments on the European network, such as the installing of standard beacons called "Eurobalise" and GSM-R deployment. Indeed, the most complete version of ETCS relies heavily on the use of GSM-R. Three levels of deployment are then scheduled in order to progressively equip the railway network.

In the first level "ETCS level 1", the trackside equipment transmits information to the train in order that it calculates its maximum authorized speed. The information given by the trackside signalling (lights and traffic signs allowing the driver to know the permitted speed), can be forwarded to the train by the Eurobalise beacons located along the track.

The second level "ETCS level 2" includes a partial deployment of the GSM-R and information can then be forwarded to the train by the GSM-R. The position of trains is still detected by trackside systems but the trackside signalling is no longer necessary since all information is transmitted directly to the train.

Finally, the third level aims to optimise railway lines capacity and further reduce the trackside equipment. ETCS Level 3 is a major revision of the classic management system which is based on fixed intervals between the trains. In ETCS level 3, the route is thus no longer managed in fixed track sections but the intervals depend on the braking distances. The trains find their position themselves by means of positioning beacons or sensors and transmit the positioning signal to the radio block centre.

Then, this highlight the GSM-R is an essential and safety component in the management of the railway European network and it is necessary to warrant its immunity facing the railway electromagnetic environment (Midya, 2008).

4. Control of the radiated EM emissions in railway

The railway environment is a severe electromagnetic environment where railway equipment performs safety critical functions. Additionally, the railway runs very close to commercial and residential areas. For these reasons, it is important to provide guidance on EMC issues by applying specific EMC standards to railway applications. These standards fall generally into two categories: governmental standards, such as the EN50121:2006 part 1-5 (EN50121, 2006) published by European Committee for Electrotechnical Standardization

(CENELEC), describing EMC for railway applications, or railway industry standards such as Railtrack Group Standard GM/RC 1031 (GMRC1500, 1994), which provide guidance on EMC between railway infrastructure and trains.

A complete list of standards related to railway applications is presented and discussed in (Konefal et al., 2002), some of these standards are presented in the table 1 for convenience.

EN 50121 parts 1-5	Railway Applications Electromagnetic Compatibility
CISPR/C/116/CDV	Interference from overhead power lines, high voltage equipment and electric traction systems.
GM/RC 1500	Code of Practice for EMC between the Railway and its Neighbourhood
EN 55011 (CISPR 11)	ISM radio-frequency equipment – Radio disturbance characteristics – (CISPR 11) Limits and methods of measurement
UMTA-MA-06-0153-85-6	Conductive Interference in Rapid Transit Signalling Systems, Suggested Test Procedures for Conducted Emission Test Vehicle
UMTA-MA-06-0153-85-8	Inductive Interference in Rapid Transit Signalling Systems Suggested Test Procedures for Inductive Emissions of Vehicular Electrical Power Subsystem, Rail-to-Rail Voltage from 20 Hz to 20 kHz
UMTA-MA-06-0153-85-11	Radiated Interference in Rapid Transit Signalling Systems Suggested Test Procedures for Broadband Emissions of Rapid Transit Vehicles -140 kHz to 400 MHz

Table 1. List of EMC standards applied to railway domain

The standards applied in Europe in order to characterize the EM environment in the railway context are the EN 50121 while in USA, the electromagnetic emission limits are imposed by the Urban Mass Transportation Administration of the U.S. Department of Transport (UMTA standards). The EN 50121 standards notably aim to control the emission levels from the railway infrastructures to the outside world while UMTA standards aim to avoid interferences with the wayside equipment (transit signalling systems). In both cases, no method is proposed to characterize the EM environment on board trains, i.e. above, inside, and under the trains, especially where new and future sensitive systems can be located.

The standards EN 50121 indicate the methodologies and the limits to apply, relating to the EM emissions and immunity of railway equipment, vehicles and infrastructures. The emissions of the whole railway system, including vehicles and infrastructure are dealt with the section 2 of the EN 50121. The objective of the tests specified in this standard is to verify that the EM emissions produced by the railway systems do not disturb the neighbouring equipment and systems. The methodology then consists in measuring the radiated EM emissions at a distance of 10 m from the middle of the tracks and at about 1.5 m from the ground and to compare them with the maximum allowed levels. The limits are specified for the frequencies included between 9 kHz and 1 GHz. The measurement protocol is specified for four frequency bands which are 9 kHz-150 kHz, 150 kHz - 30 MHz, 30 MHz - 300 MHz and 300 MHz-1 GHz.

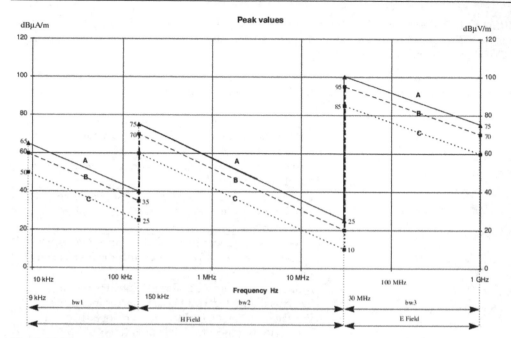

Fig. 1. Emission limits according EN 50121-2. A=25 kV ac; B = 15 kV ac, 3 kV dc or 1.5 kV dc; C = 750 V dc and bw1 = 200 Hz; bw2 = 9 kHz; bw3 = 120 kHz

The basis for the level derived in EN50121 has been the actual levels measured at a number of railways sites around Europe. While the scope of this standard covers the frequency range DC to 400 GHz, in practice limits are not set above 1 GHz. In general words, this standard does not consider the wider impact on the radio spectrum, it mostly sets the actual stage of the current levels around the railway structure.

Additionally, when comparing the EN 50121 standards to common EMC measurement standards, it is noted that there are several crucial differences in the methods of measurement. In many EMC tests, emission limits are specified in terms of a measurement with a quasi-peak detector (QP). However, the use of a quasi-peak detector in EN 50121 standards is not possible due to the highly dynamic environment. For EN 50121, a peak detector is prescribed.

5. EM immunity of the railway equipment and systems

The railway immunity levels for radiated interference are comparable with those specified by the industrial generic standard; 10V/m for trackside equipment. For rail borne equipment mounted externally to the rolling stock or within the driver's cab or passenger compartment 20V/m is specified. This level is comparable to the 24V/m specified by the Automotive Directive 95/54/EC. However, tests are different from automotive industry where full vehicle tests are performed in anechoic chamber to guarantee that all the systems can work together in the presence of electromagnetic disturbances. In the railway environment full vehicle tests are often not feasible. Due to the dimension and the speed of the trains, they cannot be tested in nominal operating condition inside an anechoic chamber. In this context, component and sub-system testing becomes very important to prevent EMC problems.

Although the immunity levels presented in EN 50121 provide an overall view of the railway electromagnetic environment, they are not suitable to perform immunity tests on the on board components, especially in the case of modern communication systems such as GSM-R. Additionally, high speed trains as other rolling stock apparatus are supplied by a catenary. In this particular context the train can be considered as a fixed equipment supplied by an electrical network. Consequently, EMC standard EN 61000-4-4 should apply. This standard aims at defining a common and reproducible basis for the evaluation of the performances of electrical and electronic equipment facing electrical fast transients on its different inputs. It is clearly adapted to test the immunity of the electronics and we would have referred to it if our objective had been to test the electronics of a GSM-R mobile.

However, as it will be shown in section 7, the test signals defined in this standard EN 61000-4-4 differ significantly from the typical transient disturbances received by GSM-R antennas. Additionally, as presented in (Knobloch, 2002), modern communication systems use digitally coded radio signals that operate with a much smaller signal-to-noise ratio (SNR) in comparison to analogical ones. The explanation lies in the fact that digital data streams are discontinuous and include redundancy to correct errors. (Knobloch, 2002) also points out that peak detector or QP detectors are not suitable to convert electromagnetic disturbance in some measure of deterioration in communication. Consequently, it is important to envisage component immunity testing solution which permits us to evaluate the telecommunication system against electromagnetic conditions representative of the railway electromagnetic environment.

6. The GSM-R communication system

The GSM-R (Global System for Mobile communication-Railways) is a wireless digital communication system, based on the public European GSM Phase 2+. This system is used to ensure the vocal exchanges and to transmit railway signalling information between trains and railway control centres. The GMS-R is currently deployed in numerous European countries in order to ensure the interoperability of trains throughout the whole European railroad network. The GSM-R system includes two parts:

-   dedicated base stations, called Base Transceiver Station (BTS) installed along the railway tracks, and connected to railroad control centres, through a wired network.
-   GSM-R antennas installed on the roof of train locomotives and connected through shielded cables to GSM-R mobile on board the train, as shown in Fig. 2.

The base stations are generally spaced from about 3 or 4 km and the GSM-R signal level has to be superior to -92 dBm, 95 % of the time and the space (UIC, 2003). In practise, the power of the received signal on board train varies between -20 dBm at proximity of the base station and -90 dBm at middle distance between two successive base stations (Hammi, 2009).

Catenary

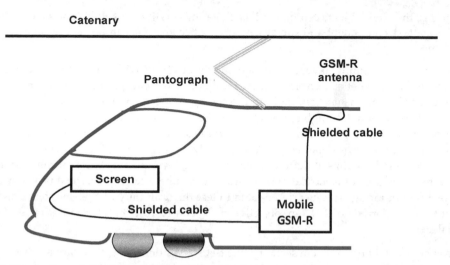

Fig. 2. Illustration of the on-board GSM-R system

The GSM-R is used in order to maintain a continuous voice and data link between the train and the control centres, and different trains located in the same neighbourhood. In the final version of ERTMS, the train sends its position through the uplink (from the train to the base stations) and receives signalling traffic information (speed limit, pass-through authorization…) through the downlink (from base stations to the train).

In Europe, the GSM-R uplink occupies the frequencies between 876 MHz and 880 MHz and the downlink between 921 MHz and 925 MHz. These frequency bands are separated by a frequency bandwidth dedicated to public and extended GSM.

Each frequency band used by the GSM-R is divided into 19 frequency channels of 200 kHz bandwidth. Only 19 channels are used by the system, in order to reduce the risk of interference with the public and extended GSM, using adjacent frequency bands.

The GSM-R is a TDMA (Time Division Multiple Access) system. The information is transmitted through each channel, according to 4.516 ms periodical TDMA frames, divided into 8 time intervals called "Time Slots" of 577 µs. During this time slot, the transmitted information is called burst, including 156 bytes, transmitted during 3.7 µs.

The data transmitted through the GSM-R system are very sensitive and the good operation of the GSM-R system is crucial to the capability and security of the European railway network. Thus, this system has been developed in order to be robust, with the capability of standing to some electromagnetic interference (EMI).

In fact, the GSM-R is included in the Euroradio protocol, which is specific to the railway and manages with altered received information, notably by resending some altered bursts until good reception. The use of such robust communication system is essentially motivated by the severity of the railway electromagnetic environment and the safety requirements.

In the next section we will focus on the different EMI that the GSM-R transmission signals can meet in the railway electromagnetic environment.

7. The EM noise sources affecting the GSM-R signals

On board a moving train and in normal operating conditions, the GSM-R system can meet different transient or permanent EMIs, with various amplitudes, time durations, repetition rates, frequency bands... Moreover, the GSM-R antennas are generally multi-band antennas and are not really selective around the frequency bands dedicated to the railway. They can thus receive GSM-R in-band and out-band EMIs (Mansson, 2008).

(Mansson, 2008) showed that out-band EMIs observed in railway environment could be a serious threat to the low noise amplifier (LNA) installed at the GSM-R receiver input. The susceptibility of this component can be reached with such EMIs and permanent damages on the system can happen.

In this effort, we will mainly focus on the in-band EMIs acting basically on the GSM-R useful signal. A description of the sources and different characteristics of these disturbances will be presented in the next part. Their impact on the GSM-R communication will also be described.

7.1 Description of the EM noise sources

From an EMC point of view, the railway infrastructure is a harsh complex EM environment where cohabitation between high power and digital communication systems with numerous eventual coupling mechanisms could be hazardous for the useful signal of the GSM-R. In this part, we will show that, on board moving trains, the GSM-R system is mainly affected by transient EM disturbances occurring between the catenary and the pantograph, in addition to the permanent disturbances coming from the public GSM base stations.

Fig. 3 synthesizes the different EMIs that could impact the GSM-R useful signals and describes the mechanism responsible of the generation of the transient disturbances on a GSM-R antenna fixed on the roof of a train. In fact, when a bad sliding contact occurs between the catenary and pantograph, a transient event could appear between these elements. This phenomenon can be observed with naked eye as a spark appearing between the catenary and the pantograph. Thus, a transient current circulates through these elements, which behave as transmission antennas, emitting EMI that the GSM-R antenna can receive.

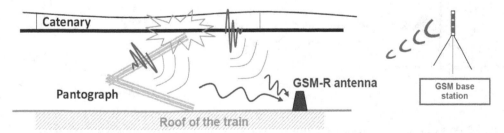

Fig. 3. EMIs received by GSM-R antenna and acting on the GSM-R useful signals

The generated wideband signal can easily cover the frequency bandwidth of the GSM-R system. However, from the train side, the GSM-R transmissions are mainly vulnerable to the EMIs covering the down-link frequency band. Indeed, on board trains, the signals emitted by the GSM-R antenna (up-link) have power levels highly superior to the power levels of the useful signals received by the antennas (down-link).

In addition, the GSM-R system uses frequency bands quietly close to the public GSM bandwidths, and when public GSM base stations use the adjacent frequency bands of the GSM-R, the risk for the GSM-R communications increases. This phenomenon is mainly observed when the train is operating in the vicinity of a city, where public GSM base stations and user numbers highly increase.

7.1.1 Transient EMI acting on the GSM-R useful signal

Measurement campaigns carried out on board moving trains (Hammi, 2009) showed that the transient events, triggered when a bad sliding contact occurs between the catenary and the pantograph, are the most penalizing events for the GSM-R useful signals. Fig. 4 (a) shows an example of a transient signal recorded by an oscilloscope connected to a GSM-R antenna. The analysis of a large number of transients collected on board trains showed that their time duration is generally inferior to 20 ns (Ben Slimen, 2009), with a typical value of 5 ns and a typical value of the rise time is 0.4 ns. Fig. 4 (b) shows the maximal EM amplitude generated by 284 successive transient events on the downlink frequency band of the GSM-R in normal operation conditions. Each point in this graph links the rank of the recorded transient and its maximal amplitude within the 921 – 925 MHz frequency band, corresponding to the down-link frequency band.

These results show that these transients generate high level EMIs that can reach - 40 dBm. Moreover, statistical analysis (Ben Slimen, 2009) of these transient disturbances highlighted that they can be very frequent, especially on high speed lines.

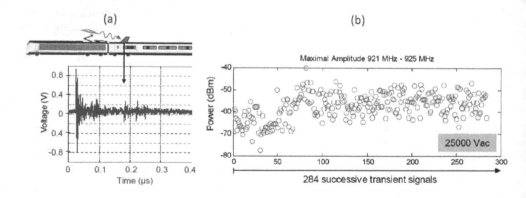

Fig. 4. (a) Example of transient disturbance in time domain and (b) maximal EM power generated by 284 successive transients in downlink GSM-R band

7.1.2 Permanent EMI acting on GSM-R antenna

Measurement campaigns have been carried out on board moving train, on railway lines equipped with the GSM-R system, in order to show the relation between the public GSM signal and the permanent interferences that can be observed in the closest downlink GSM-R frequency channel to the public GSM band (Hammi, 2009).

In Fig. 5 the top curve gives the variation of the EM power signal obtained through the last GSM-R channel along 20 km. The second one is the result obtained into an intermediary channel, supposed to be free. The last curve is obtained through the first public GSM frequency channel.

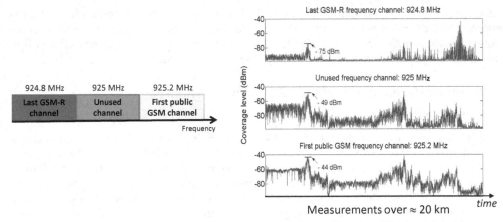

Fig. 5. EM measurement in different frequency channels used by the GSM-R and public GSM along 20 km railway line equipped with GSM-R system

Fig. 5 shows clearly that the variation of the measured amplitudes into the unused frequency channel and the last GSM-R frequency channel are partially similar to the EM noise variation observed through the first public GSM channel. This result proves that the public GSM signal sent through the first channel can disturb the GSM-R bandwidth, and confirms that the public GSM signal can be considered as a serious threat to the GSM-R useful signal. Indeed, as can be seen on Fig. 5, a signal of -44 dBm on the first public GSM channel induces an EMI of -75 dBm on the last GSM-R frequency channel. Knowing that the GSM-R signal level can decrease to a minimum value of -92 dBm (UIC, 2003), this interference level can be sufficient to severely disturb a GSM-R transmission on this channel.

7.2 Impacts of the EM noises

According to the type of the disturbance affecting the GSM-R system, the effect of the received EMI can vary. In fact, the transient events taking place between the catenary and the pantograph are wideband disturbances that can affect all the frequency channels used by the GSM-R system. It is obvious that the useful signal sent through the GSM-R system at the occurrence of the transient disturbance will be somehow disturbed. In fact, compared to the 3.7 µs time duration of one GSM-R bit, the transient duration of some ns is quietly small. So we need to investigate the real impact of such short time duration events on the interpretation of a disturbed GSM-R bit. This work will be presented in the next part of this effort.

When it comes to permanent EM disturbances coming from the public GSM base stations, the impact of these disturbances can be mainly observed when the GSM-R system is using the last frequency channel and the public GSM system is using the first frequency channel. So, even if the last frequency band of the GSM-R system is affected by the public GSM signal, the useful GSM-R signal is not necessary affected. In fact, when such scenario occurs, the GSM-R system could use different advanced protocols, mainly Euroradio protocol that can allow the system to stand to such disturbances. However, when the whole GSM-R channels are used, the system could be really affected by these disturbances.

However, among the current EMC standards, none methodology of immunity testing is adapted to the GSM-R system and to the characteristics of the EMIs that it can meet in the railway environment. The next section is then focused on a specific immunity approach to this system in order to assess the real risks for the GSM-R transmissions.

8. Immunity testing of the GSM-R signals in laboratory

8.1 Methodology

To evaluate the impact of the interferences on the quality of the GSM-R transmissions, a GSM-R mobile was employed. This mobile can be connected to one network either simulated by a specific piece of equipment or coming from a base station installed at proximity. In our case, we use a network simulator allowing controlling the network parameters such as GSM-R channel used for the communication, power of signals... and performing BER measurements. The principle of this methodology of test is first connect the mobile to the simulated network and to establish a communication with the simulator. Then, the EM disturbance signals (permanent noise + transient signals) were generated and their impact on the quality of the GSM-R communication was evaluated thanks to criteria introduced in the following paragraph. Fig. 6 gives an illustration of this test methodology.

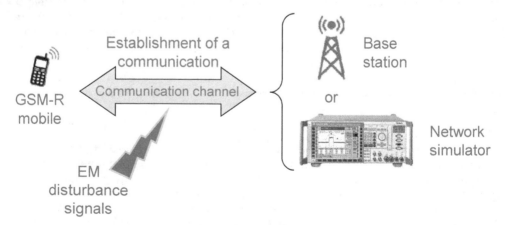

Fig. 6. Principle of the employed methodology of test

8.2 Definition of the immunity criteria

Two immunity criteria can be employed: Rxqual and Bit Error Rate (BER). The BER corresponds to the percentage of erroneous bits in a given transmission length (Breed, 2003):

$$BER = \frac{Number\ of\ erroneous\ bits}{Total\ number\ of\ bits} \times 100\% \tag{1}$$

The Rxqual is a quality parameter measured by the GSM-R mobile and it defines the quality of the received signal on a level from 0 to 7 (the lower the Rxqual is, the higher is the quality). The Rxqual is linked to the BER calculated on the learning sequences included in the GSM-R frames. The specifications defined in standards such as (ITU-T, 2003) require that the Rxqual is inferior or equal to 3 in order to ensure a good quality of communication. A relationship exists between BER and Rxqual: each value of Rxqual is associated with a range of values of BER (Lagrange et al., 1996) as can be seen in the following table.

RXQUAL	BER	
Quality level	Range of values	Typical value
0	BER < 0.2 %	0.14 %
1	0.2 % < BER < 0.4 %	0.28 %
2	0.4 % < BER < 0.8 %	0.57 %
3	0.8 % < BER < 1.6 %	1.13 %
4	1.6 % < BER < 3.2 %	2.26 %
5	3.2 % < BER < 6.4 %	4.53 %
6	6.4 % < BER < 12.8 %	9.05 %
7	12.8 % < BER	18.10 %

Table 2. Correspondence table between BER and Rxqual

8.3 Employed test bench

The test bench which was employed to perform immunity tests in laboratory is presented in Fig. 7. This test bench aims at reproducing the EM conditions that the GSM-R system is susceptible to meet on board trains. It is composed of three main parts:

- the communication system which consists in a GSM-R mobile connected to a network simulator called CMU 200 from Rohde & Schwarz.
- the EM noise generation which permits us, thanks to the two signal generators, to simulate the presence of permanent and EM transient noises simultaneously or separately.
- the area "analysis in frequency domain" is used to control the power of the exchanged signals at the input of the GSM- R mobile.

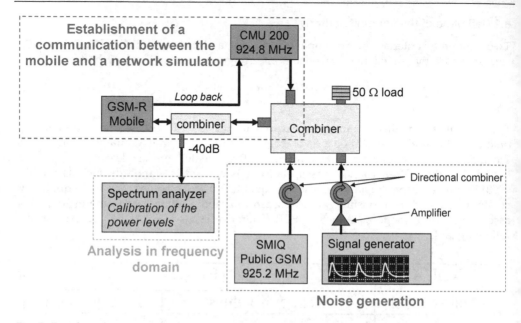

Fig. 7. Employed immunity test bench

8.4 Employed test signals

The GSM-R communication is established using the last useful GSM-R channel (924.8 MHz) from the down-link frequency band. The power level of the signals generated by the network simulator is adjusted so as to obtain a level of -70 dBm at the input of the GSM-R mobile. That corresponds to realistic operational conditions on board trains.

As for the public GSM signals, the communication channel employed is the first one (925.2 MHz) which is adjacent to the last useful GSM-R channel (924.8 MHz) used for the tests. The level of these signals is variable in order to study the effect produced on the quality of the GSM-R communication depending on the power level of the interference signals.

The signal used to simulate the presence of transient signals is a double exponential (duration=5ns, rise time=0.4 ns) modulated by a sinus at the frequency 923 MHz which corresponds to the center frequency of the GSM-R down-link frequency band. The corresponding mathematical expression is the following one:

$$S(t) = A \times (e^{-\frac{1}{D}t} - e^{-\frac{1}{RT}t}) \times u(t) \times \sin(2\pi Ft) \qquad (2)$$

where D=5 ns, RT=0.4 ns, F=923 MHz and u is the unit step function.

The values employed for rise time (RT) and duration (D) result from a statistical analysis we performed on transients collected on board trains during one measurement campaign (Ben Slimen, 2009). Fig. 8 gives the time representation of this test signal.

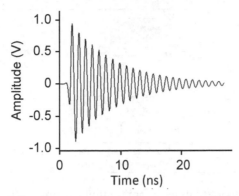

Fig. 8. Used transient test signal

Contrary to duration and rise time, it is not possible to determine a typical value for the recurrence of transients since it is very variable and depends on several operating conditions (speed of the train, one or two pantographs, state and age of the catenary and pantograph...). During the measurement campaign performed on board trains, we generally noticed that very few transients appeared at low speed whereas they could occur with a time interval of about 5 µs at about 200 km/h. As a consequence, the recurrence of transients is considered as a variable parameter for the immunity tests: for each measurement, the transient disturbances are generated with a constant time interval (TI) between two successive transients as illustrated in Fig. 9 and the immunity results are given in relation to the value of the time interval.

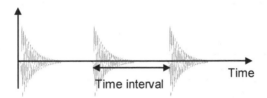

Fig. 9. Illustration of the time interval (TI) between the successive transient disturbances

9. Results of EM immunity tests on GSM-R transmissions

9.1 Configurations of test

Three different configurations, as shown on Fig. 10, are considered when studying the effect produced by the interference signals on the quality of the GSM-R transmissions:

-   presence of public GSM signals only,
-   presence of transient signals only,
-   presence of public GSM and transient signals simultaneously.

The aim is, in a first step, to observe and quantify the impact of each type of interference separately and in different conditions of test (different power levels for permanent interferences, different time intervals for transient interferences...). In a second step, the combined effect of the two types of disturbances is assessed.

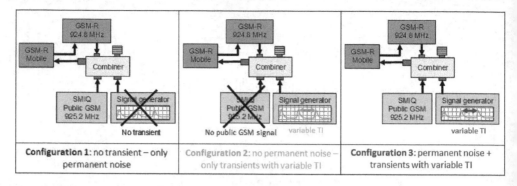

Fig. 10. The different configurations used for immunity testing

9.2 Impact of public GSM signals

The configuration of test corresponds to "configuration 1" from Fig. 10. As previously explained in paragraph 8.4, the GSM-R transmissions takes place on the channel 924.8 MHz with a power level of -70 dBm and the public GSM ones on the channel 925.2 MHz with a variable level from -72 to -12 dBm. Fig. 11 presents the results of the BER measurements as a function of the power level of public GSM signals. The vertical axis gives the value of the BER in % and the horizontal axis represents the power level of the public GSM band signal (on the channel 925.2 MHz) which induces the permanent noise on the GSM-R channel.

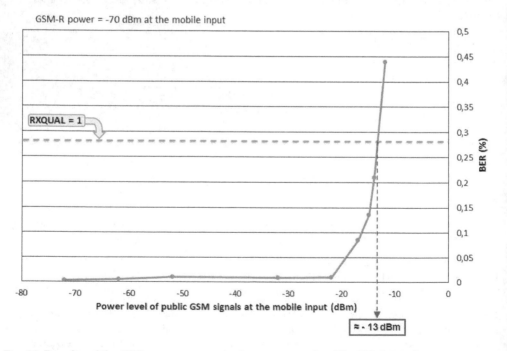

Fig. 11. Results of the BER measurement in the presence of public GSM signals

In Fig. 11, the public GSM signal has to exceed -20 dBm to start affecting the quality of the GSM-R communication (the BER starts to increase). That means that the interference signals on the 400 kHz adjacent channel have to be 50 dB higher than the wanted signal on the GSM-R communication channel to deteriorate the quality of the transmission, which well complies with the specifications (ETSI, 2000). Indeed, the standard EN 300 910 stipulates that a mobile has to tolerate a 400 kHz adjacent interference level of -50 dB.

Then, we also notice that a level of -13 dBm is necessary to induce a Rxqual equal to 1. It will be highlighted later that this level is different when transient interferences are simultaneously present.

9.3 Impact of transient EM interferences produced by catenary-pantograph sliding contact

These tests and measurements are related to "configuration 2" in Fig. 10. The GSM-R signal can be set to the desired value and the interference level produced by transient signals on the GSM-R frequency band can be controlled by using a variable attenuator in order to obtain the desired signal-to-noise ratio (SNR) at the mobile input. As for the measurements, during a test sequence we vary the time interval between two consecutive transients and one measure of BER is made for each chosen time interval. Then, the same test sequence is applied with one other signal-to-noise ratio. Three different SNR at the mobile input are tested: +5, 0 and -5 dB. The results are presented in Fig. 12 where the vertical axis of the graph corresponds to the value of the BER in % and the horizontal axis gives the time interval between two successive transients in µs.

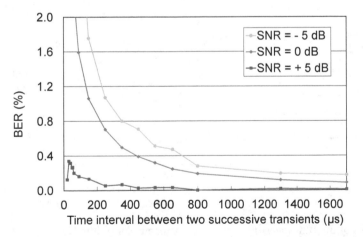

Fig. 12. Results of the BER measurement in the presence of transients for different values of the signal-to-noise ratio (SNR)

The first thing to notice is that the BER evolves with the time interval between transients: it increases with the recurrence of transients. Indeed, the BER is higher for small values of time interval whatever the value of the SNR. In (Adriano, 2008), a relation was proposed to estimate the BER from the TI between the transient interferences, under the assumption that the SNR is equal to 0 dB.

The second thing to observe is that the SNR has an impact on the BER. Indeed, if taking the curve obtained for SNR = 0 dB (at the mobile input) as a reference, we see that, when the transient level is 5 dB higher than the GSM-R signal (SNR=-5 dB), the measured BER increases. Consequently, the transmission could be more severely disturbed when the SNR decreases to -5 dB whereas, in the reverse case (SNR=+5 dB), the BER is lower (less than 0.4 %) which guarantees a good quality of communication whatever the recurrence of the transient interferences.

9.4 Tests and measurements in the presence of both types of interference signals simultaneously

In this section, we now consider "configuration 3" in Fig. 10: presence of permanent noise and transients simultaneously with two arbitrarily chosen values for the transient time interval which are TI=150 µs and TI=550 µs. The following graph, on the right of Fig. 13, shows the results of the BER measurements in this configuration of test. The first curve (black one with points) corresponds to the evolution of the BER without transient and the two others (orange with squares and blue with triangles ones) with transients for the two considered values of time interval. These values were chosen so that 3 transients can occur during the time duration of one GSM-R burst in the first case (TI=150 µs) and only one in the second case (TI=550 µs), as can be seen in the illustration on the left of Fig. 13.

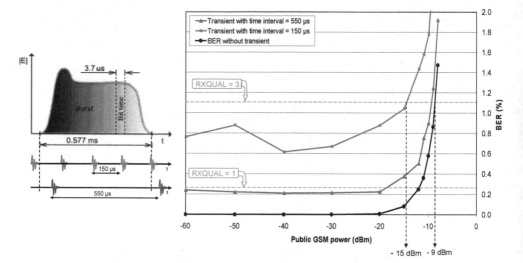

Fig. 13. Results of the BER measurements in the presence of public GSM signals and transient signals with GSM-R signal power = -70 dBm at the mobile input

In the absence of transient signals (black curve with points), the public GSM signals have to reach a power level of -9 dBm to induce a Rxqual equal to 3 whereas in the presence of transient disturbances with a time interval of 150 µs, a level of -15 dBm is sufficient. In other words, the impact on the GSM-R communication of the transient disturbances "adds" to the one of signals in the public GSM band. We thus conclude that the susceptibility of the GSM-R to permanent noise is higher in the presence of transient disturbances.

Obviously, these results are linked to the GSM-R signal power used for the test (-70 dBm) and we would obtain a better level of immunity if setting up the GSM-R signal to a higher level of power. However, we are not going to develop this point in this chapter, since further studies on the immunity of the GSM-R system can be found in (Dudoyer et al., to be published).

10. Conclusion

This chapter outlined the major developments underway on the European rail network and highlighted the electromagnetic vulnerability of the GSM-R which is a key component of the management system. Indeed, immunity testing carried out in laboratory to confront the GSM-R transmissions to EM disturbances representative to those measured on trains, have shown that the quality of the transmissions can be significantly affected. The results of the section 9 highlighted that the impact of the transient disturbances on the quality of the GSM-R transmissions is linked to two main factors: the levels of noise produced on the GSM-R down-link frequency band and the repetition rate of the transient disturbances. Moreover, their impact can also be related to the presence of permanent interferences with the GSM public. Consequently, the assess of the risks of disturbances of the GSM-R transmissions requires to monitor the spectral distribution of the EM noise over the time, and with a high temporal resolution which permits us to perform direct comparison with the transmission of the digital data.

The current European standard methodologies of measurement of the EM emissions in the railway domain (EN 50121, 2006) which only consist in spectral analysis of the radiated emissions without taking into account the time dimension are then not adapted to the control of the EM emissions in order to protect the GSM-R transmissions. This chapter which proposed a methodology to perform immunity testing of GSM-R transmissions in laboratory, has also highlighted the main features of the EM noise it is necessary to characterize on board trains to ensure that the radiated emissions will not affect the ability of the GSM-R system.

11. Acknowledgment

The authors of this chapter would like to thanks SNCB and SNCF to have given them access to their trains to perform measurements in real conditions and also ALSTOM which provided them specific railway equipment. This work was performed in the framework of the RAILCOM project supported by the PCRD 6 and CISIT projects supported by the North Region and the FEDER.

12. References

Adriano, R., Ben Slimen, N., Deniau, V., Berbineau, M. & Massy, P. (2008). Prediction of the BER on the GSM-R communications provided by the EM transient disturbances in the railway environment, *Proceeding of EMC Europe*, pp. 771-775, Hamburg, Germany, September 2008.
Ben Slimen, N., Deniau, V., Rioult, J., Dudoyer S. & Baranowski S. (2009). Statistical characterisation of the EM interferences acting on GSM-R antennas fixed above

moving train. *The European Physical Journal Applied Physics,* Vol. 48, No.2, (2009), ISSN 1286-0042

Breed, G. (2003). Bit Error Rate: Fundamental Concepts and Measurement Issues. *High Frequency Electronics,* Vol.2, No.1, (January 2003), pp. 46-48

Dudoyer, S., Deniau, V., Adriano, R., Ben Slimen, M. N., Rioult, J., Meyniel, B. & Berbineau, M. (to be published). Study of the Susceptibility of the GSM-R Communications to the Electromagnetic Interferences of the Rail Environment, *accepted for publication in IEEE Transactions on electromagnetic compatibility.*

EN 50121. (2006). European standards for Railway applications: Electromagnetic Compatibility-Part 1-5

ETSI EN 300 910 V8.5.1 (2000). European Standard on Digital cellular telecommunication system (Phase 2+); Radio transmission and reception, November 2000.

GM/RC 1500, (1994). Code of Practice for EMC between the Railway and its Neighbourhood, December 1994

Hammi, T., Ben Slimen, N., Deniau, V., Rioult, J. & Dudoyer, S. (2009). Comparison between GSM-R coverage level and EM noise level in railway environment, *Proceedings of the 9th International Conference on Intelligent Transport System-Telecommunications, ITST 2009,* pp. 123-128, Lille, France, October 2009

IEC 60050-161. (1990). International Electrotechnical Vocabulary. Chapter 161: Electromagnetic compatibility, Edition: 1.0, International Electrotechnical Commission, 1990

ITU-T Recommendation K.48 (2003). EMC requirements for each telecommunication equipment - Product family Recommendation, July 2003.

Jarašūnienė, A. (2005). General Description of European Railway Traffic Management System (ERTMS) and Strategy of ERTMS implementation in Various Railway Managements, *Transport and Telecommunication,* Vol.6, No.5, (2005), pp. 21-27, ISSN 1407-6160

Knobloch, A. & Garbe, H. (2002). Critical Review of Converting Spectral Data into Prospective Bit Error Rates, *Proceeding of IEEE International Symposium on Electromagnetic Compatibility, Volume 1,* pp. 173-178, ISBN 0-7803-7264-6, Minneapolis, MN, USA, August 2002

Konefal, T., Pearce, D.A.J., Marshman, C.A. & McCormack, L.M., (2002). Potential electromagnetic interference to radio services from railways, *Final Report, University of York,* 2002, Retrieved from <http://www.yorkemc.co.uk/research/railways/>.

Lagrange, X., Godlewski, P. & Tabbane, S. (1996). GSM-DCS Networks, Hermès, Paris, 1996, pp. 207-208, ISBN 2-86601-558-4

Mansson, D., Thottappillil, R., Bäckström, M & Lundén, O. (2008). Vulnerability of European Rail Traffic Management System to Radiated Intentional EMI. *IEEE Transactions on Electromagnetic Compatibility,* Vol.50, No.1, (2008), pp. 101-109, ISSN 0018-9375

Midya, S. & Thottappillil, R. (2008). An Overview of Electromagnetic Compatibility Challenges in European Rail Traffic Management System, *Journal of Transportation Research Part C: Emerging Technologies,* Elsevier, (Ed.), Vol.16C, Issue.5, (2008), pp. 515-534, ISSN 0968090X

UIC Project EIRENE (2003). System Requirements Specification, v14.0, 21 October 2003. Source : EIRENE Project Team

Permissions

The contributors of this book come from diverse backgrounds, making this book a truly international effort. This book will bring forth new frontiers with its revolutionizing research information and detailed analysis of the nascent developments around the world.

We would like to thank Xavier Perpinya, for lending his expertise to make the book truly unique. He has played a crucial role in the development of this book. Without his invaluable contribution this book wouldn't have been possible. He has made vital efforts to compile up to date information on the varied aspects of this subject to make this book a valuable addition to the collection of many professionals and students.

This book was conceptualized with the vision of imparting up-to-date information and advanced data in this field. To ensure the same, a matchless editorial board was set up. Every individual on the board went through rigorous rounds of assessment to prove their worth. After which they invested a large part of their time researching and compiling the most relevant data for our readers. Conferences and sessions were held from time to time between the editorial board and the contributing authors to present the data in the most comprehensible form. The editorial team has worked tirelessly to provide valuable and valid information to help people across the globe.

Every chapter published in this book has been scrutinized by our experts. Their significance has been extensively debated. The topics covered herein carry significant findings which will fuel the growth of the discipline. They may even be implemented as practical applications or may be referred to as a beginning point for another development. Chapters in this book were first published by InTech; hereby published with permission under the Creative Commons Attribution License or equivalent.

The editorial board has been involved in producing this book since its inception. They have spent rigorous hours researching and exploring the diverse topics which have resulted in the successful publishing of this book. They have passed on their knowledge of decades through this book. To expedite this challenging task, the publisher supported the team at every step. A small team of assistant editors was also appointed to further simplify the editing procedure and attain best results for the readers.

Our editorial team has been hand-picked from every corner of the world. Their multi-ethnicity adds dynamic inputs to the discussions which result in innovative outcomes. These outcomes are then further discussed with the researchers and contributors who give their valuable feedback and opinion regarding the same. The feedback is then collaborated with the researches and they are edited in a comprehensive manner to aid the understanding of the subject.

Apart from the editorial board, the designing team has also invested a significant amount of their time in understanding the subject and creating the most relevant covers. They scrutinized every image to scout for the most suitable representation of the subject and create an appropriate cover for the book.

The publishing team has been involved in this book since its early stages. They were actively engaged in every process, be it collecting the data, connecting with the contributors or procuring relevant information. The team has been an ardent support to the editorial, designing and production team. Their endless efforts to recruit the best for this project, has resulted in the accomplishment of this book. They are a veteran in the field of academics and their pool of knowledge is as vast as their experience in printing. Their expertise and guidance has proved useful at every step. Their uncompromising quality standards have made this book an exceptional effort. Their encouragement from time to time has been an inspiration for everyone.

The publisher and the editorial board hope that this book will prove to be a valuable piece of knowledge for researchers, students, practitioners and scholars across the globe.

List of Contributors

Inmculada Gallego, Santos Sánchez-Cambronero and Ana Rivas
University of Castilla-La Mancha, Spain

Anatoly Levchenkov, Mikhail Gorobetz and Andrew Mor-Yaroslavtsev
Riga Technical University, Latvia

Petzek Edward and Radu Băncilă
"Politehnica" University of Timişoara & SSF-RO Ltd., Romania

Sebastiaan Meijer
Delft University of Technology, The Netherlands
Royal Institute of Technology, Stockholm, Sweden

Clavel Edith, Meunier Gérard, Bellon Marc and Frugier Didier
G2Elab Grenoble Electrical Engineering Laboratory, Saint Martin d'Hères, BP46, Saint Martin d'Hères, Cedex, France

Shodolapo Oluyemi Franklin
University of Botswana, Gaborone, Botswana

Gbenga Matthew Ayininuola
University of Ibadan, Ibadan, Nigeria

More Ramulu
Central Institute of Mining & Fuel Research, Nagpur, India

Stephen Dudoyer and Virginie Deniau
Univ Lille Nord de France, IFSTTAR, France

Nedim Ben Slimen
Assystem, France

Ricardo Adriano
Federal University of Minas Gerais, Brazil

Printed in the USA
CPSIA information can be obtained
at www.ICGtesting.com
JSHW011358221024
72173JS00003B/337